CW00375684

GUIDE

THE
GRESLEY
LEGACY

A CELEBRATION OF INNOVATION

THIS ONE'S FOR PETER HERRING.

Thanks for all the help, patience and encouragement.

STEAM CLASSIC GUIDE

THE GRESLEY LEGACY

A CELEBRATION OF INNOVATION

MARTIN SMITH

ARGUS BOOKS

Argus Books
Argus House
Boundary Way
Hemel Hempstead
Herts HP2 7ST
England

First published by Argus Books 1992

ISBN 1 85486 079 8

Publisher's Note

Although some of the older timetable extracts are not
of the best quality they have been included for their
historical interest.

Phototypesetting by The Studio, Exeter
Printed and bound in Great Britain by Clays Ltd., St Ives plc, Bungay

Contents

Introduction

Of all the celebrated locomotive designers in British railway history, unquestionably the most famous name is that of Nigel Gresley. Admittedly, Gresley had a distinct advantage when it came to securing a place at the head of the hall of fame, and that was due to the exploits of his A4 4-6-2 *Mallard* which, on 3 July 1938, charged down Stoke Bank between Grantham and Peterborough at 126 mph. That world speed record for steam haulage still stands.

The record-breaking locomotive was famed not only for its speed but also for its looks. It was a member of the class which had been the first fully-streamlined engines in Britain and, although a few purists were underwhelmed by their appearances, the machines captured the public's imagination. From that moment on, Nigel Gresley could do little wrong.

Gresley's railway career started in 1893 when he left school at the age of seventeen. In 1911, he secured the position of locomotive engineer for the Great Northern Railway at the salary of £1,800 per annum: a respectable sum in those days. At the grouping in 1923, the London & North Eastern Railway was spoiled for choice as to whom to appoint as chief mechanical engineer because the company's constituents included the North Eastern, Great Central and Great Eastern Railways. Their respective superintendents were Sir Vincent Raven, John Robinson and Alfred Hill, all of whom were highly accomplished and respected engineers but, as those three eminent gentlemen were all approaching retirement age, Nigel Gresley was the obvious choice for long-term continuity of locomotive development. The LNER was not to regret its decision.

There was considerable speculation throughout the LNER's network that Gresley would construct vast quantities of Great Northern-style locomotives, and thereby dispense with the designs which had been favoured by his pre-grouping contemporaries, but the fears were largely unfounded. Although several ex-Great Northern designs were perpetuated after 1923, Gresley had no hesitation in ordering more D11 class 4-4-0s in 1924 and these locomotives were pure Great Central. Furthermore, the D49 class 4-4-0s, which first appeared in 1927, incorporated a number of design features which leant, quite unmistakably, on the work of Vincent Raven of the North Eastern Railway. With the help and enthusiasm of his chief assistant Oliver Bulleid, Gresley experimented with design features which had appeared on railway locomotives elsewhere in Britain and also in Europe and America. Like many masters of their respective crafts, Gresley was always willing to learn from others and even the external design work of his celebrated A4s was inspired by something he had seen in France.

Life was not all sweetness and light for Nigel Gresley but, considering that his experimental work often broke new ground in the field of railway engineering, it was hardly surprising that some of his locomotives were less than earth-shattering. His P1 and, later, P2 2-8-2s were impressive looking and extremely powerful machines but they proved to be, quite simply, too big for their full potential to be realised. Gresley's most spectacular flop was the revolutionary W1 class 4-6-4 No 10000, which showed that a locomotive could have an eye-catching appearance and total unreliability in direct inverse proportions. But, by the time No 10000 was withdrawn for complete rebuilding, Gresley was basking in the limelight once again as a result of the introduction of his A4s.

Gresley received his knighthood in July 1936 but, within a couple of years, his close colleagues started to notice a decline in his health. Nevertheless, he continued his work with enthusiasm and, even when he was taken ill shortly after the public debut of his newly-designed V4 class 2-6-2 No 3401 in 1941, a speedy recovery was anticipated. To the shock of all, Sir Nigel Gresley passed away on 5 April 1941 at his home in Hertford. Had he survived until 30 September, he would have reached the retirement age of sixty-five.

The legacy which Gresley left the LNER and, subsequently, British Railways, included some of the finest locomotives in the country. Apart from his famous A4s, his other 4-6-2s were highly impressive machines but, among the railway fraternity, it is often considered that his best all-round design was that of the less celebrated V2 2-6-2s. Gresley's successor was Edward Thompson whose own ideas were, quite often, diametrically opposed to those of Gresley. In an attempt to establish his own reputation, Thompson dropped many tried and tested engineering features which Gresley had perfected and, furthermore, he rebuilt a number of Gresley's locomotives to fit in with his own ideals of perfection. Thompson did not, however, have either the bravery or suicidal tendencies to attempt to rebuild the A4s.

Gresley's legacy can also be judged by the lifespan of his designs. His first locomotives for the Great Northern, the J6 0-6-0s, were, admittedly, based on an Ivatt design but they were all approaching fifty years old when eventually retired. Not one single Gresley locomotive was withdrawn before his death and, apart from an A4 which was destroyed in an air raid on York in 1942, the two P1 2-8-2s of 1925 and a pair of twenty-eight year old O1 2-8-0s, every locomotive he designed survived to see Nationalisation. It is a remarkable testimony.

On a purely personal note, my own introduction to the ultimate in Gresleyana might ring a few bells with those who were not brought up in or around the old LNER system. When I first became interested in railways at the very beginning of the 1960s, steam locomotives still out-numbered diesels severalfold. Living in Bristol, I was well used to classy ex-GWR designs and, on school Railway Society trips to 'foreign' parts, I got to see many impressive machines which had their origins with companies other than my pet GWR. My favouritism took a jolt one Sunday in November 1962 when, during a particularly lengthy Railway Society trip, the coach deposited my colleagues and me outside New England shed in Peterborough. Much to my immense delight, the shed was full of Gresley classics. The V2s looked good and the A3s looked even better but, just inside the shed was not one but three of the A4s which, until that moment, I had only read about or studied in photographs. I distinctly remember how, on my return home, I asked my parents about the chances of our moving to Cambridgeshire.

This book does not attempt to chart the histories of every one of the locomotives, or even classes, which were designed by Nigel Gresley. During the thirty years for which he was responsible for the motive power of the Great Northern Railway and the LNER, he introduced or substantially modified around forty different classes and such an output is, of course, beyond the scope of a single volume. This book concentrates on the stories of the locomotives which were developed from the late 1920s onwards, the period when Nigel Gresley was at his peak and when his most famous and innovative machines appeared. It was during those years that he introduced the greatest ingredients of his remarkable legacy.

MARTIN SMITH
COLEFORD
SOMERSET

CHAPTER 1

Warming Up

It has been suggested that the railway career of Herbert Nigel Gresley was inspired by the Great Western Railway's William Dean. Between the ages of fourteen and seventeen, Gresley studied at Marlborough College and, just down the road at Savernake, was the GWR's main line to south-west Wiltshire and beyond. Master Gresley was a frequent visitor to the line and many of the expresses passing through were hauled by Dean's celebrated 2-2-2s. Not content with just ticking off the relevant numbers in his Combined Volume, young Nigel started to entertain thoughts of working with the machines.

At Marlborough College, the prospect of a career with railway locomotives most decidedly ruled out the job of engine driver. The college was a popular place of education for sons of the clergy, and the atmosphere was more of gentility than axle grease. On leaving school, Gresley was anxious to obtain an apprenticeship in railway engineering and he was accepted by the London & North Western Railway and placed in the care of the human dynamo, Francis Webb.

Gresley had been born on 19 June 1876 in Edinburgh, a stronghold of the North British and Caledonian Railways. His traditional family home was at Netherseal in Derbyshire, not far from Midland Railway territory and, during his school days, he had enthusiastically watched activities on the GWR. When the L&NWR offered the apprenticeship, it could be assumed that Gresley's railway affections were divided in as many as five ways but there was another company, one with which Gresley had had no protracted contact, which held a magical appeal for him. That was the Great Northern Railway.

From the L&NWR, Gresley joined the Lancashire & Yorkshire Railway and he sped through the ranks at an impressive rate so that, by the age of twenty-eight, he was the assistant superintendent of the Carriage and Wagon Department at Newton Heath. In 1905, one year after Gresley had secured the assistant superintendent's post, the position of chief carriage and wagon superintendent on Gresley's beloved Great Northern Railway had become vacant and, clutching an impressive CV and with a glowing reference from the L&Y's John Aspinall, Gresley got his job at Doncaster.

Having secured one of the top jobs on the GNR, it was hardly surprising that there was not the same scope for rapid promotion as there had been on the L&Y but, when Henry Ivatt retired in 1911, the door was open for Gresley to take the final step to the top. The GNR gave Gresley the position of chief mechanical engineer on 1 October 1911 and the appointment was of mutual benefit. On one hand, the company secured the services of what eventually turned out to be one of the greatest designers of them all and, in return, Gresley was provided with a very solid base from which to work because locomotive development on the GNR had previously had the benefit not just of Henry Ivatt but also his predecessors, Patrick Stirling and Archibald Sturrock. Gresley inherited a sizeable stud of passenger locomotives but, in general, the engines were not in the 'heavy haulage' league. In common with the Midland Railway, the GNR had a policy of running lightly-loaded passenger trains and, even on the important Anglo-Scottish expresses, weights seldom exceeded 300 tons. In the speed stakes, however, the GNR had been in the forefront in the late nineteenth century when Stirling's singles did battle in the 'Races to the North' in 1888 and 1885. During Ivatt's incumbency, 4-4-0s had started to appear in numbers after 1896 and, in 1898, Doncaster had turned out Britain's first 4-4-2, GNR No. 990. The Atlantics had not proved to be the heaven-sent answer to all ills and, when additional locomotives were required in 1900 to haul lightly-loaded fast trains to Yorkshire in competition with the Midland Railway, Ivatt returned to the tried and tested idea of 4-2-2s. The final Ivatt single, No 270, emerged from Doncaster in September 1901 and went down in the annals of railway history as the last single-driver to be built in Britain.

At the start of Gresley's term of office, the 2-2-2s and 4-2-2s were running express services to and from Kings Cross. The fastest 156-mile run to Doncaster was scheduled for 165 minutes, while an evening train from Grantham was timetabled to cover the 105 miles to Kings Cross in 110 minutes. However, it was rare for either of those services to have a loading in excess of four coaches. By contrast, the 2pm express from Wakefield to Kings Cross was often burdened with the stupendous load of six coaches and, in order to cover the 180 miles in the allotted 187 minutes, an Ivatt Atlantic was required. The schedules were definitely fast but the light loadings were treated with contempt by some of the GNR's competitors which, in general, regarded trains of less than twelve coaches as namby-pamby affairs.

In view of the lightweight nature of GNR express trains, Gresley did not attach any immediacy to the development of new passenger locomotives but the

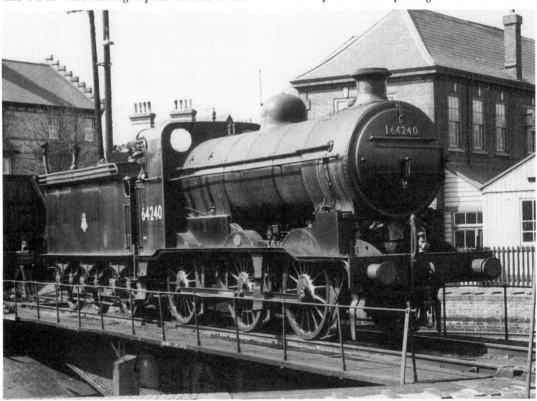

Gresley's first order for the GNR was for a class of 0-6-0s which were, basically, a continuation of a tried and tested Ivatt design, albeit with Gresley modifications. The locomotives were to become the LNER's J6 class and most had lifespans of well over forty years. GNR No 591 was built in October 1914 and became LNER No 3591 after the grouping; in 1946, it became No 4240. This picture shows it as BR No 64240 on the turntable at Cambridge shed on 14 April 1952. Its good condition was a result of a recent works overhaul and the tender, loving care of the staff at its home depot of Hitchin.

Photo: E. H. Sawford

situation regarding goods engines was a little more desperate. The GNR had been swept up in the national trend for speedier freight services but the company had an embarrassing lack of fast goods locomotives. For a while, the most unlikely engines had to be commandeered for freight expresses and, as if the sight of Ivatt 4-4-2s and 4-4-0s on freight workings wasn't unorthodox enough, the use of Stirling 2-4-0s and 2-2-2s raised a few eyebrows.

Gresley considered that the simple answer to the goods locomotive situation was to continue building Ivatt's excellent 0-6-0s, and so the first engines to be ordered from Doncaster after Gresley had taken charge could not, despite incorporating several of the new lad's own features, realistically be credited to him. The usefulness of Ivatt's basic design can be gauged by the numbers built and also their longevity as, by the end of 1922, a total of ninety-five similar locomotives had been constructed to Gresley's orders. The GNR classified these 0-6-0s as J22 and they later became the J6 class of the LNER. A batch of ten Ivatt-designed 0-6-0s was ordered by Gresley in 1912 but these had 5ft 8in coupled wheels instead of the 5ft 2in

diameter of their predecessors. These went on to become the LNER's J2 class and nine survived until the early 1950s.

The first locomotive that can be attributed wholly to the design capabilities of Nigel Gresley was GNR No 1630, which went into action on 3 October 1912 at the head of the 5.10pm Kings Cross to Baldock passenger train. No 1630 was Gresley's prototype 2-6-0 and it had been completed at Doncaster works some six weeks before its public debut. Although it made its grand entrance on a passenger working, No 1630 had initially been designed to augment the 5ft 8in 0-6-0s on express freight duties but, when the GNR's traffic department came to grasp that such freight workings were done mainly at night, it dawned on them that the engine would be available for secondary passenger duties during the daytime. After extensive tests of No 1630, nine similar 2-6-0s were built between February and April 1913 and these took Nos 1631–39. They had 5ft 8in driving wheels, 20in × 26in cylinders and, with a boiler pressure of 180lb, a commendable tractive effort of 23,400lb resulted. Although their potential as mixed traffic

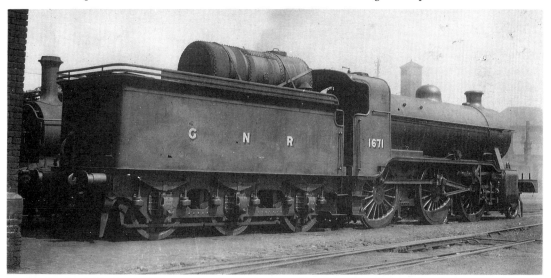

The first Gresley 2-6-0s for the GNR were the K1s which appeared in 1912–13. A larger boilered version, the K2s, first materialised in 1914 and No 1671 was one of the 1918 batch which was built by the North British Locomotive Co. Originally, the order for the batch had been placed with Beyer Peacock but wartime conditions had placed such a strain on that company that the order had had to be transferred. The cab of No 1671 shows that, even in Gresley's day, Patrick Stirling's influence had not totally disappeared from Doncaster; also evident in the photograph is the oil-burning equipment which was carried from June to November 1921 as a precaution against coal shortages due to a miners' strike. No 1671 eventually became BR No 61761 and was retired in January 1961.

Photo: H. C. Casserley Collection

During the early 1930s, many of the K2s finally had their Stirling-style cabs replaced by side-window versions. This was good news for the crews at Eastfield shed in Glasgow where the K2s were regularly used on the somewhat draughty run to Fort William and Mallaig. It had been August 1924 when the first K2 had been transferred to the former-North British section and, by 1932, twenty members of the class were resident north of the border. Here, LNER No 1787 *Loch Quoich* is seen leaving Fort William with the morning goods for Mallaig in August 1947. The following year, the engine was given BR No 61787.

Photo: Rail Archive Stephenson

engines had been seen, their intended purpose as fast freight engines was not forgotten, and six of them were allocated to Colwick for working braked goods trains to Liverpool and Manchester.

Gresley developed a larger diameter boiler in 1913 and he decided to use this for his next 2-6-0s. In April 1914, the first of the new-style 2-6-0s emerged from Doncaster and, by September 1921, sixty-three similar engines had been delivered. During World War One, the 2-6-0s were found to be invaluable, not only because of their ability to handle the fast goods trains, but also because they released the versatile J2 and J6 0-6-0s for general duties. Gresley's final development of the 2-6-0 for the GNR was the three-cylinder version, ten of which were completed between March 1920 and August 1921. The GNR designated the 2-6-0 classes as H2, H3 and H4 in the order of their appearances and the LNER reclassified them as K1, K2 and K3. The K1s were all eventually rebuilt as K2s, and the

final representative of the first true Gresley design, GNR No 1666, was retired from Kings Cross shed as BR No 61756 in June 1962.

Apart from locomotives for the fast freight services, Gresley knew that motive power at the heavyweight end of the range warranted revision. From 1901, the heaviest goods trains on the GNR had been hauled by Ivatt's 0-8-0s, fifty-five of which had been delivered by the end of 1909. The GNR's freight traffic increased beyond all expectations and, despite the power of the 0-8-0s and the gradual fitting of superheaters, the heavy loadings meant that the locomotives were being worked to capacity far more regularly than was healthy for them. After his arrival in the hot-seat, Gresley tackled the problem by designing a 2-8-0, the theory behind this wheel arrangement being that a pony truck would help to support a more powerful front end.

The first 2-8-0s were completed in 1913. They

In Scotland, the largest allocation of K3s was always at Eastfield shed and, perhaps, possessiveness was an Eastfield trait as the cleanest part of No 61789 *Loch Laidon* seems to be its 65A shedplate. The picture was taken on 26 August 1955.

Photo: E. H. Sawford

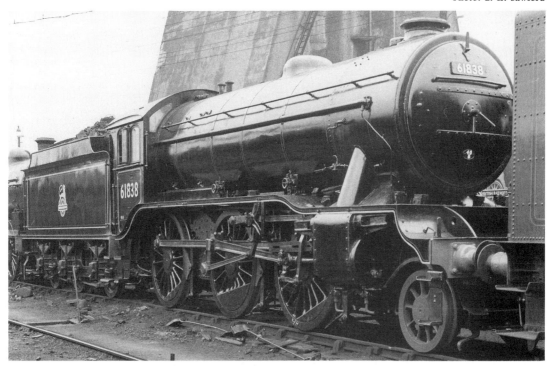

The K3 class was Gresley's three-cylinder development of his previous 2-6-0 designs for the Great Northern Railway. The class was considerably expanded by the LNER after the grouping and No 127 was built at Darlington in January 1925. Here, it is seen as BR No 61838 fresh from the works at Doncaster on 24 June 1956 before being dispatched to his home shed of Immingham.

Photo: E. H. Sawford

incorporated boilers which were larger than on any engine previously built at Doncaster and, from new, were fitted with 24-element Robinson super-heaters. With 4ft 8in coupled wheels and two 21in × 28in cylinders, the tractive effort of the new creations was an impressive 31,000lb but only five were constructed at first. Eventually, in 1919, fifteen similar 2-8-0s were delivered but, the previous year, an almost identical engine had been built, not with two cylinders but three. The increase in power which was offered by the three-cylinder 2-8-0 was deemed adequate for the foreseeable future and so, when the next batch of 2-8-0s was built in 1921, they were also three-cylinder versions.

Although Gresley had unquestionable ability and foresight, he was not averse to absorbing ideas from other engineers both in this country and abroad. During his days on the L&NWR at Crewe, his mentor, Francis Webb, had displayed what

some called bravery and others called cockiness when it came to experimenting with cylinder arrangements. Webb had been a keen proponent of compounding although, with the benefit of hindsight, it can be said that the only truly successful compounds to be used in Britain were the Midland Railways' 4-4-0s and the GWR's 4-4-2s. The Midlands compounds had come, surprisingly, from the sketch pad of Samuel Johnson, whereas the GWR's had been purchased from France for evaluation by George Churchward. About the time Gresley was getting his feet under the table at Doncaster, the school of thought in the engineering world was that it took a brave or, possibly, suicidal designer to embark on a programme of compounding. The development and application of super-heaters was considered to be a far safer bet for a designer who valued his pension, but Gresley was well aware that even greater advantages had been

Gresley's first 2-8-0s for the Great Northern were constructed in 1913 and were two-cylinder locomotives. Those became the LNER's O1 class while the three-cylinder versions which first materialised in 1918 were classified as O2s. Construction of the O2s continued after the grouping and the LNER managed to impose its love of sub-classifications on them. The pre-grouping ones were designated O2/1s, the fifteen of 1923/24 were classified as O2/2s, the forty-one of 1932–43 became O2/3s while the rebuilding of seven of the pre-grouping engines between 1943 and 1962 was deemed to warrant the creation of sub-class O2/4. The locomotive in the picture, No 63942, was built in 1924 as an O2/2 but, even though the photograph was taken on 24 June 1956, the Great Northern-style cab is still conspicuous. The locomotive's neighbour in the yard at Doncaster is J50 0-6-0T No 68907 which is another Gresley-designed ex-Great Northern machine.

Photo: E. H. Sawford

found elsewhere with the use of more than two cylinders in non-compound locomotives.

It was Gresley's predecessor at Doncaster, Henry Ivatt, who had led the way with a non-compound four-cylinder design when 4-4-2 No 271 was built in 1902, but the locomotive was not a success. The GWR had set the standard for mass production of four-cylinder engines with Churchward's highly successful Star class 4-6-0s which had started emerging from Swindon in 1907. After being allowed to play with one of the GWR's Stars, the L&NWR's C. J. Bowen Cooke had designed the four-cylinder Claughton class 4-6-0s which, although useful, had not grabbed the headlines as Churchward's machines had done. The compromise of a modern three-cylinder, non-compound locomotive had first been seen in the form of James Holden's 0-10-0T for the Great Eastern in 1902, but the three-cylinder idea had not been applied to passenger locomotives until 1909, when John Robinson's experimental conversion of a 4-4-2 had appeared on the Great Central Railway. Several successful

three-cylinder freight and shunting engines had been designed for various railway companies between 1907 and 1913 but, apart from a batch of twenty 4-4-2s designed by Vincent Raven for the North Eastern Railway in 1911, the three-cylinder idea had not been applied to any significant number of passenger engines.

The GNR's prototype three-cylinder 2-8-0, No 461, which was built in 1918, was the first Gresley-designed locomotive to be constructed with that cylinder arrangement. Prior to the emergence of No 461, however, Gresley conducted experiments with three of Ivatt's 4-4-2s, one of which was converted from a two- to a four-cylinder engine and the lessons learned from this were combined with observations of the North Eastern's three-cylinder 4-4-2s, to produce the design of the 2-8-0. One major improvement which Gresley incorporated in his three-cylinder locomotive was the simplification of the valve gear. All other three-cylinder arrangements in use elsewhere incorporated three sets of valve gear, but Gresley managed to eliminate

The Ardsley 0-6-0Ts were introduced in 1913 and were Gresley's first tank engines for the GNR. Construction of the 0-6-0Ts continued until 1930 although different batches incorporated different features; LNER No 601 was one of a series built in 1926 with left-hand drive and steam brake. Eventually, the Ardsleys were classified as J50s but the differentials ensured a proliferation of sub-classes.

Photo: H. C. Casserley Collection

Gresley's N2 class 0-6-2Ts were based on Ivatt's N1s but had higher pitched boilers. However, the Gresley engines were intended to fit the Metropolitan Railway's loading gauge and so the additional height had to be offset by smaller chimneys and lower domes. The picture shows GNR No 1753 which was built in March 1921; it was retired as BR No 69532 in June 1959. The design of the N2s was so sound that construction continued until 1929; a smaller version, which had originated with Alfred Hill on the Great Eastern, became classified as N7s and Gresley ordered a quantity of these for the LNER between 1924 and 1928.

Photo: H. C. Casserley Collection

one set in his version. No 461 was put to work on heavy coal trains from Peterborough to Hornsey and, with its capability of hauling loads of 1,300 tons, it demonstrated the advantages of three-cylinder operation admirably.

The GNR's three-cylinder 2-6-0s, Nos 1000-09, also incorporated Gresley's innovative valve gear when they were built in 1920/21. These locomotives had six-foot diameter boilers and, with a pressure of 180lb, they put in some remarkable performances. During the coal strike of 1921, the GNR had to abandon its liking for lightly loaded passenger trains, and the three-cylinder 2-6-0s regularly found themselves in charge of trains consisting of twenty bogie coaches. Even with these most untypical loadings, they recorded speeds of up to 75 mph with commendable regularity.

It was not just main line requirements which concerned Gresley. In 1913/14, his first 0-6-0Ts were constructed and these were intended specifically for the steeply-graded lines in the West Riding of Yorkshire. The engines, which had 4ft 8in coupled wheels and 18in × 26in cylinders, were fitted with boilers of 4ft 8in diameter and these had, by necessity, been removed from Ivatt's 0-8-2Ts. In 1920, Gresley introduced a class of superheated 0-6-2Ts with 5ft 8in driving wheels and 19in × 26in cylinders; these were to become the well-known N2s, and the majority of the class were fitted with condensing apparatus so that they could work over the lines of the Metropolitan Railway.

When the spectre of the grouping was looming, it seemed that Gresley was determined to see the Great Northern Railway disappear with rather more than just a whimper. The six-foot diameter boilers and the three-cylinder arrangement of the 1920 2-6-0s were both incorporated in the last new Gresley design before the grouping, the product of which was 4-6-2 No 1470 *Great Northern*. The only 4-6-2 to have been previously used in Britain was the GWR's No 111 *The Great Bear* which, despite

The very first Gresley Pacific was A1 class GNR No 1470 *Great Northern* which was completed at Doncaster in April 1922. This picture of the engine was taken at Doncaster in April 1934, by which time some of the later A1s had been rebuilt as A3s. No 4470, as the engine had become in 1925, escaped rebuilding as an A3 but its reprieve was not indefinite as Edward Thompson rebuilt it to his 'new A1' specification in 1945. As BR No 60113, the locomotive survived until November 1962.

Photo: Rail Archive Stephenson

Apart from *Great Northern*, the only other Gresley Pacific which was built for the GNR was No 1471 *Sir Frederick Banbury*. Completed in July 1922, the engine was rebuilt as an A3 in October 1942; it was retired as BR No 60102 in November 1961. This picture shows it in superb condition at Doncaster in June 1958.

Photo: Rail Archive Stephenson

its inestimable publicity value for its owners, had had a strictly limited sphere of operations since its introduction in 1908.

In contrast to the GWR's monster, the GNR's 4-6-2 was designed more from necessity than prestige. As early as 1915, Gresley had toyed with the idea of an enlarged 4-4-2 which would be capable of hauling main line expresses of up to 600

LONDON, PETERBRO', NOTTINGHAM, RETFORD, MANCHESTER, LIVERPOOL, DONCASTER, LEEDS, YORK, &c.—G.N.

Offices—King's Cross Station. N.1. Gen. Man., C. H. Dent. Sec., E. Burrows.

Week Days.

[A detailed Great Northern Railway "Down" timetable is reproduced here, listing stations including KING'S CROSS, Broad Street, Finsbury Park, Hatfield, Cambridge, Hitchin, Three Counties, Arlesey and Shefford Road, Biggleswade, Sandy, Tempsford, St. Neots, Offord and Buckden, Huntingdon, Abbotts Ripton, Holme, Yaxley and Farcet, Peterbro', Cromer (Beach), Tallington, Essendine, Little Bytham, Corby, Great Ponton, Grantham, Lincoln (High Street), Nottingham, Newark, Retford, Worksop, Sheffield (Victoria), Stockport (Tiviot Dale), Manchester (Lon. Rd), Manchester (Central), Southport (Lord St.), Liverpool (Central), Barnby Moor and Sutton, Ranskill, Scrooby, Bawtry, Rossington, Doncaster, Hull (Paragon), Wakefield (Westgate), Bradford (Exchange), Wakefield (Kirkgate), Halifax, Huddersfield, Holbeck, Harrogate, Ripon, Bradford (v. Holbeck), Leeds (Central), Arksey, Moss, Balne, Heck, Temple Hirst, Selby, Riccall, Escrick, Naburn, York, Scarborough, Newcastle (Central), Berwick, Edinboro' (Waverley), Glasgow (Queen St.), Dundee (Tay Bridge), Aberdeen, Perth (v. Forth Bridge), Inverness, with numerous columns of departure and arrival times for morning and afternoon services.]

For Notes, see pages 336 and 337; for Continuation of Trains, see pages 334 to 337.

King's Cross, &c., to the North.] 332 [Great Northern Main Line.] 333

An extract of the services on the Great Northern Railway's main line is shown in this timetable for the summer of 1922. The locomotives in charge of the expresses would have been, in the main, the 4-4-2s which were designed by Gresley's predecessor, Henry Ivatt.

tons and, although both 4-6-0s and 2-6-2s were considered, the result was the 4-6-2 which was delivered in April 1922. Three months later, No 1470 was presented with a chum, No 1471, which was later named *Sir Frederick Banbury*. The 4-6-2s did all that was expected of them and this was well illustrated by No 1471 when, on a trial run on 3 September 1922, it hauled a 610-ton train over the 105 miles from Kings Cross to Grantham in a mere 122 minutes. Despite the uncertainty caused by the forthcoming grouping, the GNR ordered a further ten 4-6-2s, each with the 6ft 8in coupled wheels and three 20in × 26in cylinders as used by their class leaders, but the first of these was not delivered until February 1923, by which time the London & North Eastern Railway had taken control of matters. The 4-6-2 which was delivered in February 1923 was, incidentally, LNER No 4472 *Flying Scotsman*.

During the eleven years in which Gresley oversaw the GNR's locomotive matters, there were many indications for the future. In his first two years in

the chair, Gresley had introduced two highly useful main line designs, the 2-6-0s and the 2-8-0s, and also the Yorkshire 0-6-0Ts, later known as the Ardsley Tanks. It was not, however, just the designs themselves which were significant but also their adaptability. All three of those early designs had proved to be eminently suitable for updating and this had been illustrated by the development of three-cylinder locomotives from the original designs of the two-cylinder 2-6-0s and 2-8-0s. Construction of similar three-cylinder 2-6-0s continued until 1937 and, in the end, the class comprised 193 locomotives. The Ardsley Tanks reflected a similar story. The first had been completed in 1913, but the basic design was modified to keep pace with contemporary requirements over the years and the last of the 102 members of the class did not appear until 1939.

The longevity, as well as the designs, of Gresley's individual locomotives had also become evident early on in his career. In GNR days, Gresley had introduced nine different classes which, by the end

TABLE 1.1: SUMMARY OF GRESLEY LOCOMOTIVES FOR THE GREAT NORTHERN RAILWAY

GNR CLASS	LNER CLASS	TYPE	DRIVING WHEELS	CYLINDERS	BOILER PRESSURE	DATES BUILT	TOTAL BUILT	NOTES
A1	A1	4-6-2	6ft 8in	(3) 20 × 26	180lb SU	1922	2	(*)
H2	K1	2-6-0	5ft 8in	(2) 20 × 26	180lb SU	1912–13	10	(a)
H3	K2	2-6-0	5ft 8in	(2) 20 × 26	180lb SU	1914–21	65	(a)
H4	K3/1	2-6-0	5ft 8in	(3) 18½ × 26	180lb SU	1920–21	10	(*)
J22	J6	0-6-0	5ft 2in	(2) 19 × 26	170lb	1912–22	95	(b)
J23	J51/1	0-6-0T	4ft 8in	(2) 18½ × 26	175lb	1913–19	30	(c)(*)
J23	J50	0-6-0T	4ft 8in	(2) 18½ × 26	170lb	1922	10	(d)(*)
N2	N2	0-6-2T	5ft 8in	(2) 19 × 26	170lb SU	1920–21	60	(*)
01	01	2-8-0	4ft 8in	(2) 21 × 28	180lb SU	1913–19	20	(e)
02	02	2-8-0	4ft 8in	(3) 18 × 26	180lb SU	1918	1	
02	02/1	2-8-0	4ft 8in	(3) 18½ × 26	180lb SU	1921	10	

Notes:
(*) More of class added after the grouping.
(a) Two K1 class locos rebuilt as K2 class 1920/21. The others were similarly treated 1931–37.
(b) This design originated with H. A. Ivatt in 1909 and 35 locos were built during his term of office. Several of Ivatt's locos and most of Gresley's were superheated from new.
(c) Rebuilt and reclassified in 1929–35 as J50/1 and J50/2.
(d) Reclassified as J50/2 in 1939.
(e) Reclassified as 03 in 1944.

of 1922, had yielded 313 locomotives.

Just three of those pre-grouping locomotives failed to see Nationalisation in 1948 and only one of the classes was to become extinct before 1962. The very first Gresley-designed locomotive, GNR No 1630 of August 1913, was withdrawn without ceremony as BR No 61720 in June 1956.

The grouping, which took effect on 1 January 1923, was akin to Judgement Day for many of Britain's locomotive designers. Although the London & North Eastern Railway chose to centralise its administrative operations at Kings Cross, the main locomotive works were to remain at Doncaster and Darlington. The former guv'nors of the North Eastern and Great Eastern Railways, Vincent Raven and Alfred Hill, were joined by Nigel Gresley and the Great Central's John Robinson as the nominal contestants for the top job with the LNER. There was not, however, any catfight.

The LNER was impressed with Raven's work, particularly in the field of electrification, but it was considered that, at the age of sixty-five, he could not be expected to fend off the lure of the golf course for long. Alfred Hill of the Great Eastern was also close to retirement, and so was John Robinson, but the latter was formally offered the LNER post out of respect to his overwhelming seniority. Robinson took the anticipated course and declined the LNER's offer and so the way was left clear for the youngster of the bunch, forty-seven year-old Nigel Gresley, to become the chief mechanical engineer of the LNER.

The grouping manifested itself in different ways on each of the big four. To the Great Western, it was little more than the absorption of small but, nevertheless, highly useful Welsh companies. The Southern Railway split neatly into three geographical sections, the boundaries of which were not a world

TABLE 1.2: LNER AND BR RENUMBERING OF PRE-1923 GRESLEY LOCOMOTIVES

LNER CLASS	FIRST LNER NOS.	1946 NOS.	BR NOS.	LAST WITHDRAWAL/S	
				LOCO/S	DATE
A1 (a)	4470–4471	113/102	60113/60102	60113	11/62
J6	3536–3610 3621–3640	4185–4279	64185–64279	64203/26/77	6/62
J50 (b)	3221–3230	8920–8929	68920–68929	68922 (c)	9/63 (c)
J51/1 (d)	3157–64/66–76/78 3211–3220	8890–8919	68890–68919	68892/904/08 (e)	9/63 (e)
N2	4606–4615 4721–4770	9490–9549	69490–69549	69504/20/23/ 29/35/38/46	9/62
K1 (f)	4630–4639	1720–1729	61720–61729	61728	12/60
K2 (f)	4640–4704	1730–1794	61730–61794	61756	6/62
K3/1	4000–4009	1800–1809	61800–61809	61807	11/62
O1	3456–3460 3462–3476	3475–3494	63475–63494 (g)	63484	10/52
O2	3461	3921	63921	63921	5/48
O2/1	3477–3486	3922–3931	63922–63931	63924	11/63

Notes:

(a) Locomotives reclassified: No 4470 became Thompson A1 class in 1945, No 4471 became A3 class in 1942.

(b) Redesignated J50/2 in 1939.

(c) BR No 68928 continued in Departmental use until May 1965.

(d) Redesignated J50/1 and J50/2 after rebuilding of locos in 1929–35.

(e) BR Nos 68911/14/17 continued in Departmental use until May 1965.

(f) Two K1s rebuilt in 1920/21 as K2, the others in 1931–37.

(g) Four locos withdrawn without carrying BR number.

At the grouping, the LNER inherited twenty-one of Robinson's excellent Great Central 4-6-2Ts which were subsequently classified as A5s. After the grouping, Gresley had no hesitation in ordering a further twenty-three and ten were built at Gorton in 1923 while the others were supplied by Hawthorn Leslie in 1925/26. Gresley's A5s had modifications such as redesigned cylinders and piston valves, but a number of slight alterations had to be made to the external dimensions as the ex-GNR loading gauge was not quite as generous as that of the old GCR. These two pictures both show Hawthorn Leslie-built A5s of 1925. LNER No 1760 was photographed at Darlington Bank Top in August 1937 and, at Nationalisation, became BR No 69835. The one with the well-polished tanks, No 69838, it seen on its home shed at Stockton in August 1956. Nos 69835/38 were both withdrawn in November 1958.

Both Photos: Rail Archive Stephenson

away from the demarcation lines of the pre-grouping constituents. There was a different story on the LMS where several ferociously independent companies became uneasy bedfellows, and this could also have become the case on the LNER where five sizeable and successful railways, each with its fair share of corporate ego, were thrown together. In locomotive matters, the painless accession of Nigel Gresley to the post of CME went a long way to prevent potential trouble spots.

Gresley's open-mindedness was demonstrated in 1924 when the subject of new locomotives for Scotland came up for discussion. Rather than stick with his own designs, Gresley considered that the 4-4-0s of the Great Central's improved Director class were a better bet for duties on former North British routes, and so twenty-four were ordered for delivery to depots in Glasgow, Edinburgh, Perth and Dundee. There were suspicions that the introduction of an English design to Scotland would be just the start of the spread of standard types throughout the LNER system, but the gossips were wrong. There were, naturally, some locomotive exchanges within the network but, in the main, Gresley felt that his pre-grouping contemporaries had known their stuff, and so the vast majority of pre-grouping designs were left on their respective home territories.

The production of Great Northern-designed 4-6-2s continued after the grouping and, by July 1925, a total of fifty-two was in action on express duties. Because of the urgency which was attached to building the 4-6-2s, twenty of the locomotives which were built in 1924 were constructed by the North British Locomotive Company in Glasgow. A pair of Raven 4-6-2s had been completed at the North Eastern Railway's Darlington works just weeks before the grouping, and three similar locomotives were built there in March 1924. The sheer size of the Raven engines presented a problem as, even on the main line duties to which they were restricted, they were hampered by sharp curves at Newcastle and Morpeth but, despite this, they continued in service until replaced by A4s in the mid-1930s. Under the LNER classification system, the Gresley 4-6-2s became A1s and the Raven locomotives A2s.

For lighter passenger and fast freight traffic, the delivery of further K3 class three-cylinder 2-6-0s commenced in August 1924 and, by December 1925, the class comprised seventy locomotives. They were affectionately known as 'Jazzers', partly because of their rhythmic exhaust beat but also because of a distinctive little gyration of their backsides. The sixty which were delivered in 1924/25 were constructed at Darlington and divided between the Scottish, North Eastern and Southern sections of the LNER. The original ten, which had been built in pre-grouping days, still retained their old allocations of Kings Cross, New England and Doncaster.

The LNER's inheritance of heavy freight locomotives included modern eight-coupled machines from the Great Northern, North Eastern and Great Central Railways, and so it was 1925 before Gresley had the need to think about new goods designs for the LNER. Although there was not a pressing need for additional freight engines, the Railway Centenary Exhibition of July 1925 provided the excuse for coming up with something special and the result was a pair of three-cylinder 2-8-2s. Classified P1, they carried Nos 2393/94 and, with 5ft 2in coupled wheels, 20in × 26in cylinders and a boiler pressure of 180lb, were credited with a macho tractive effort of 38,500lb. Each of the P1s had the benefit of a booster which was connected to the trailing axle in order to provide additional power and, although Gresley had previously conducted a number of experiments with boosters, it was a bold step to incorporate them on a completely new design.

In service, the P1s proved perfectly capable of hauling 1,600-ton coal trains from Peterborough to Hornsey. The length of these trains, however, was one hundred wagons and this created problems for the operating staff. Trains of that length occupied three block sections on the running track and, furthermore, passing loops of adequate length were not over-abundant. Even on arrival at Hornsey, their troubles were not over, as it was necessary to halt the trains on the running lines while shunting tanks divided them into manageable sections. The two P1s were allowed to cause disruption until July 1945 when they claimed the unwanted distinction of being the first Gresley locomotives to be deliberately withdrawn.

Only one of the P1s, No 2393, was actually exhibited at Darlington in 1925 but it was accompanied

The sheer size of Gresley's P1 2-8-2s is evident in this picture of No 2394. The two members of the class spent virtually all their lives on coal workings from Peterborough to Ferme Park and were given refreshment at Hornsey shed before making their return journeys. Here, No 2394 is seen at Hornsey on 17 June 1932.

Photo: LCGB/Ken Nunn Collection

When brand new, P1 class 2-8-2 No 2393 was exhibited at the Stockton & Darlington Centenary Exhibition. Its size impressed the masses but, when put to work, the engine was, if anything, too powerful for its own good. Like its solitary classmate, No 2393 sampled little more than the heady delights of heavy coal workings from Peterborough to Ferme Park and this picture shows it performing its customary task near Potters Bar on 11 May 1935.

Photo: LCGB/Ken Nunn Collection

The LNER was adamant that its massive Garratt 2-8-8-2T should be ready for the Stockton & Darlington Centenary Exhibition in 1925 and, such was the haste, that there was no time to treat the locomotive to the full LNER livery. Consequently, the monster appeared at the Exhibition in workshop grey and this explains the mysterious 'lighter' paintwork seen in this picture.

Photo: H. C. Casserley Collection

by another new heavy freight locomotive. This was Beyer-Garratt 2-8-8-2T, LNER No 2395, which claimed the honour of having a tractive effort of 72,940lb, the greatest of any steam locomotive ever to be used in Britain. The huge six-cylinder contraption was built specifically for banking duties behind trains of up to 1,000 tons on the 1 in 40 incline of Worsborough Bank, between Wath and Penistone in West Yorkshire. No 2395 was the first Beyer-Garratt to be constructed for a British main-line railway company and it performed its designated duties, despite the best efforts of the local water supply, until electrification rendered it redundant in 1949. It was subsequently transferred to the Lickey Incline, on the former Midland line south of Birmingham, but failed to impress local crews and was withdrawn in 1955.

Slightly further down the scale, five T1 class 4-8-0Ts were built in 1925 for heavy shunting but this design had originated with Wilson Worsdell on the North Eastern Railway in 1909. In 1926, Gresley identified a need for further freight loco-motives and designed two new classes of 0-6-0s, the J38s and J39s. Both types of 0-6-0 were very similar in a number of ways, including their 20in × 26in cylinders and 180lb boilers, but the major difference between them was that, while the J38s had 4ft 8in coupled wheels, those of the J39s were 5ft 2in. A total of thirty-five J38s was built, but construction of the J39s continued intermit-tently until 1941 and the final tally was 289 loco-motives. The J38s all went to Scotland, principally for Fife coalfield traffic, and they went on to spend the rest of their lives in Central and Eastern Scotland. Apart from short periods, the entire class was divided between St. Margarets, Dundee, Thornton and Dunfermline sheds and the last survivors, BR Nos 65901/29 (LNER Nos 1401/42), were not retired from Thornton until April 1967. The J39s became dispersed over all of the LNER system and were seen on duties as far apart as Liverpool Street and Fort William but, despite their versatility and weight of numbers, the whole lot went to the torch between 1959 and 1962.

On 7 March 1949, Gresley's U1 class Garratt 2-8-8-2T was transferred to the Midland Region for banking duties on the Lickey Incline. As BR No 69999, it was used chimney-first as had been the practice at its former haunt, the Worsborough Incline, but the crews complained that it was difficult to judge the distance between the front of the engine and the rear of the ascending train. Consequently, the engine was taken to Birmingham and turned on the Kings Norton triangle so that it could operate bunker-first, but this didn't please the crews either. This time, their grouse was that bunker-first banking caused false readings on the water gauges in the cab. The U1 lasted on the Lickey only until 1950 but, when it returned there in 1955, it found no more favour than first time round. The locomotive was eventually retired in December 1955. This picture shows it banking on the Lickey on 4 May 1949.

Photo: H. C. Casserley Collection

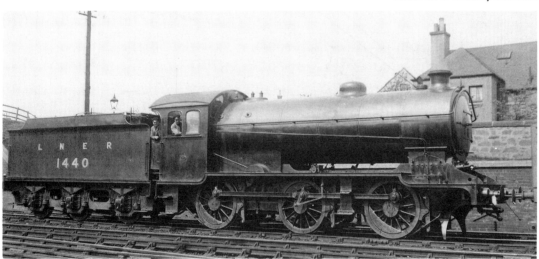

Gresley's J38s were smaller-wheeled versions of the J39s and thirty-five were built at Darlington between January and May 1926. All of the class spent their entire existences in Scotland and LNER No 1440 was retired as BR No 65927 from St. Margarets shed in December 1964.

Photo: H. C. Casserley

The J39s were Gresley's 'go-anywhere, do-anything' mixed traffic engines although they had originally been conceived as freight locomotives. Between 1926 and 1941, a remarkable total of 289 J39s were built and the one pictured, LNER No 1586, was constructed in April 1937 by Beyer Peacock. This photograph was taken at Buxton (Midland) when the engine was just a few months old. Most of the J39s survived into the 1960s and, as a class, were numbered 65700–64988. No 1586 itself became BR No 64928 and was later transferred to Malton shed in the North Eastern Area; it was retired in July 1961.

Photo: E. R. Morten

TABLE 1.3: SUMMARY OF GRESLEY CLASSES INTRODUCED 1923–1926

CLASS	TYPE	DRIVING WHEELS	CYLINDERS	BOILER LBS.	TRACTIVE EFFORT	DATES BUILT	TOTAL BUILT	NOTES
A1	4-6-2	6ft 8in	(3) 20 × 26	180 SU	29,835lb	1923–25	50	(a)
A2	4-6-2	6ft 8in	(3) 19 × 26	200 SU	29,918lb	1924	3	(b)
A5	4-6-2T	5ft 7in	(2) 20 × 26	180 SU	23,750lb	1923–26	23	(c)
J38	0-6-0	4ft 8in	(2) 20 × 26	180 SU	28,414lb	1926	35	
J39	0-6-0	5ft 2in	(2) 20 × 26	180 SU	25,664lb	1925–41	289	
J50	0-6-0T	4ft 8in	(2) 18½ × 26	170	23,636lb	1924	10	(d)
J50/3	0-6-0T	4ft 8in	(2) 18½ × 26	175	23,636lb	1926–30	38	
K3	2-6-0	5ft 8in	(3) 18½ × 26	180 SU	30,031lb	1924–37	183	(e)
N2	0-6-2T	5ft 8in	(2) 19 × 26	170 SU	19,945lb	1925–29	47	(f)
N7	0-6-2T	4ft 10in	(2) 18 × 24	180 SU	20,512lb	1924–28	122	(g)
02/2	2-8-0	4ft 8in	(3) 18½ × 26	180 SU	31,236lb	1923–24	15	(h)
P1	2-8-2	5ft 2in	(3) 20 × 26	180 SU	38,500lb	1925	2	(j)
T1	4-8-0T	4ft 7¼ in	(3) 18 × 26	175	34,080lb	1925	5	(k)
U1	2-8-8-2T	4ft 8in	(6) 18½ × 26	180 SU	72,940lb	1925	1	

TABLE 1.3 CONT.

Notes:

(a) All A1s were rebuilt as A3s between 1927 and 1948.

(b) Design originated with the North Eastern Railway for which two similar locos were built in 1922.

(c) Design originated with the Great Central Railway for which 21 locos were built between 1911 and 1917. The ex-GCR locos and those built by the LNER in 1923 were eventually redesignated A5/1 and the locos of 1925/26 became A5/2.

(d) Design originated with the Great Northern for which 10 locos were built in 1922.

(e) Design originated with the Great Northern for which 10 locos were built in 1920 and 1921. One loco was rebuilt as a K5 in 1945.

(f) Design originated with the Great Northern for which 60 locos were built in 1920 and 1921. Incorporated in N2/2 class in 1940.

(g) Includes sub-classes of N7/1, N7/2 and N7/3. Design originated with the Great Eastern for which 12 locos were built in 1914 and 1921/22. All N7 (GER locos) and N7/1s were rebuilt as N7/4 after 1940.

(h) Very similar to 02 class which originated with the Greath Northern. Further locomotives were built between 1932 and 1943 and were classified 02/3.

(j) Tractive effort with booster in operation: 47,500lb.

(k) Design originated with the North Eastern Railway for which 10 locos built in 1909/10.

TABLE 1.4: LNER AND BR RENUMBERING OF CLASSES INTRODUCED 1923–1926.

LNER CLASS	FIRST NOS.	ORIGIN	1946 NOS.	BR NOS.	FIRST WDN.	LAST WDN.
A1 (*)	2543–2582	GNR	44–83	60044–60083	1959	1966
	4473–4481		103–112	60103–60112		
A2	2402–2404	NER			1936	1937
A5/1	(r)	GCR	9820–9829	69820–69829	1958	1960
A5/2	(r)		9830–9842	69830–69842	1957	1958
J38	1400–1447 (a)		5900–5934	65900–65934	1962	1967
J39	(r)		4700–4988	64700–64988	1959	1962
J50 (*)	3231–3240	GNR	8930–8939	68930–68939	1959	1963
J50/3	(r)		8940–8977	68940–68977	1959	1963 (d)
K3	(r)	GNR	1810–1992	61810–61992	1959	1962
N2	892–897	GNR	9550–9596	69550–69596	1956	1962
	2583–2594					
	2662–2690					
N7 (*)	(r)	GER	9612–9733	69612–69733	1958	1962
02/2	487–501		3932–3946	63932–63946	1962	1963
P1	2393/2394				1945	1945
T1	1656–1660		9918–9922	69918–69922	1955	1961
U1	2395		9999	69999	1955	1955

Notes:

(*) Rebuilt and/or reclassified.

(a) Not all of the numbers in the sequence were used.

(d) Locos in Departmental use until May 1965.

(r) Random numbers.

During the mid-1920s, the increasing commuter traffic in the London area meant that additional locomotives were required for fast suburban traffic and Gresley felt that he would be pushed to improve on tried and tested designs. For services out of Kings Cross, he ordered the construction of further N2 0-6-2Ts and a total of forty-seven were delivered between 1925 and 1929. Several of the N2s found their way to Scotland, West Yorkshire and the former Great Central system but, nevertheless, they went on to become synonymous with Kings Cross until the 1960s. For suburban services over the former Great Eastern network into Liverpool Street, Gresley stuck with the N7 class 0-6-2Ts which had been designed by Alfred Hill in 1914. The N7s had 4ft 10in coupled wheels instead of the 5ft 8in of the N2s but, otherwise, the two classes were not dis-similar. Between 1925 and 1928, a total of 112 more N7s were built but, much to the disgust of ex-Great Eastern staff, none of them were constructed at Stratford.

By the end of 1926, Nigel Gresley was still only fifty years-old but had notched up over fifteen years as a chief mechanical engineer. His four years with the LNER had already shown evidence that the characteristically long shelf lives of his designs was being continued as his 2-6-0s and 4-6-2s were as fresh as when they had first appeared in pre-grouping days. Gresley had, by this time, laid the solid foundations from which a whole series of classic designs was to materialise. Developments at Doncaster up to the end of 1926 had been pretty impressive but Gresley still had rather a lot left to show the world.

CHAPTER 2

From Shires to 'The Galloping Sausage'

The first of the D49 engines were the Shires and the class leader, LNER No 234 *Yorkshire*, materialised in October 1927. This picture shows it as BR No 62700 at York on 8 September 1951.

Photo: E. R. Morten

Until 1927, the only passenger locomotives built for the LNER had been constructed to pre-grouping designs. The A1 4-6-2s and K3 2-6-0s, although highly acclaimed, had originated with the Great Northern Railway. The family tree of the A2s harked back to the North Eastern Railway and the D11 4-4-0s, which Gresley had despatched to the former North British Railway territory in Scotland, had a pedigree that was pure Great Central. When the original member of the D49 class 4-4-0s emerged from Darlington works in 1927, it went down in the records as the first LNER-designed passenger

locomotive but, ironically, the D49s were to be last type of LNER 4-4-0s to be built. The need for the D49s arose indirectly from the effects of the grouping.

On 1 January 1923, the LNER had acquired two-fifths of Scotland. Of the pre-grouping Scottish companies, the Caledonian, the Highland and the Glasgow & South Western Railways had joined the LMS camp, while the LNER had absorbed the North British and the Great North of Scotland Railways. At one time, there had been plans for a fifth post-grouping company which would look after matters north of the border and, had it not been for in-fighting among the railway clans, the 'big four' of 1923 might well have been the 'big five'.

In terms of locomotive ownership, the LNER's inheritance from Scotland had consisted of the largest and the smallest of the pre-grouping Scottish railways. By close of play in 1922, the North British had contributed 1,075 locomotives to the LNER's lists while the Great North of Scotland had added just 122. Unlike the Scottish companies, which passed to Derby control, neither of the LNER's Scottish constituents had turned to 4-6-0s, let alone 4-6-2s, for passenger duties. The Great North of Scotland Railway had perpetuated the tradition of 'the longer the name, the smaller the railway' and had had little need for anything larger than 4-4-0s and, at the end of 1922, these accounted for 100 of its 122 locomotives. The North British Railway had accumulated a stock of twenty-two 4-4-2s but these were greatly outnumbered by 4-4-0s. On the Great North, all but sixteen of the 4-4-0s had been built before 1900 but, although the North British

had not been without its share of veterans, a healthy number of its four-coupled passenger locomotives had been very modern and efficient machines.

The insistence of the North British for persevering with four-coupled locomotives for its express services had been largely down to the nature of its lines. The Waverley route from Edinburgh to Carlisle might have been famed for its scenery and history, but it was the cause of regular nightmares in the company's operating department. It had taken a strong argument from William Reid to introduce 4-4-2s on the line and neither he nor his successor, Walter Chalmers, were going to stick their necks out and push for six-coupled passenger engines. When 4-6-2s became available after the grouping, it was unsurprising that they found the Waverley route to be decreed out of bounds but, disappointingly, the routes to Aberdeen and Dundee were also placed off limits. The versatile D11s of 1924 helped out to a great extent in Scotland but, by 1926, a need was identified for upgraded 4-4-0s

This photograph of D49/2 No 247 *The Blankney* was taken at Scarborough shortly after the locomotive had been built in July 1932 and the immaculate fully-lined green livery emphasises just how stylish the D49s were. When new, this engine was allocated to Leeds and it remained in the North Eastern Area throughout its life; at Nationalisation, it became BR No 62741 and was withdrawn from Botanic Gardens shed in October 1958. The hunt after which it was named was based around Lincoln and Newark.

Photo: E. R. Morten

which would be powerful enough to augment the LNER's assembled stud of 4-4-2s on express duties in Scotland. The outcome was Gresley's D49s.

Drawings for the D49s were prepared early in 1926 but it was October 1927 before the first, No 234 *Yorkshire* came on the scene. By June 1929, thirty-six D49s were in service and twenty-three of these were despatched to Scotland, the remainder being allocated to the former North Eastern Railway area. The D49s had 6ft 8in coupled wheels and their boilers and fireboxes were virtually identical to those which had been developed earlier for the J38 and J39 0-6-0s. The new arrivals used Gresley's favoured three-cylinder arrangement although, when they were in the planning stage, it had been intended to build two of them as compounds.

Gresley demonstrated his absence of megalomania by incorporating on the D49s several features which smacked decidedly of Vincent Raven, and these included the use of a single casting for the cylinders and steam chests, inside steam-pipes and 3ft 1¼in bogie wheels. As if to display a smidgin of individuality, a batch of eight D49s which materialised in 1929 differed from their predecessors by having screw reversing gear and steam brakes instead of steam-operated reversers and Westinghouse braking.

Six of the D49s which were completed in 1928 were fitted with Lentz valves instead of piston valves, whereas the two which had been earmarked for compounding incorporated, instead, Lentz rotary cam gear. The rotary gear was still in the development stage when Gresley first thought of using it on the D49s, and so ex-GER J26 0-6-0 No 8280 was volunteered as a guinea pig to evaluate the equipment. The detail differences among the D49s enabled the LNER to demonstrate its penchant for confusing matters by the use of as many sub-classes as possible. The conventional locomotives became D49/1, those with Lentz rotary gear D49/2 and those with standard Lentz gear became D49/3. The classification department would have had a field day had a further variation of the D49s got past the drawing board stage; this was for a six-cylinder

TABLE 2.1: ORIGINAL DIMENSIONS OF THE D49 CLASS 4-4-0s

BUILT:	Darlington 1927–1935
WEIGHTS FULL (locomotive):	65 tons 11 cwt (a)
(tender):	52 tons 0 cwt (b)
TENDER CAPACITY (water):	4,200 gallons
(coal):	7 tons 10 cwt
WHEELBASE (locomotive):	24ft 11in
(tender):	13ft 6in
WHEEL DIAMETERS (leading):	3ft 1¼in
(coupled):	6ft 8in
CYLINDERS:	(3) 17in × 26in
BOILER PRESSURE:	180 lb
HEATING SURFACES (tubes):	871.75 sq ft
(flues):	354.53 sq ft
(firebox):	171.5 sq ft
(superheater):	271.8 sq ft
GRATE AREA:	26 sq ft
TRACTIVE EFFORT @ 85% boiler pressure:	21,556 lb
LNER ROUTE AVAILABILITY:	8
BR POWER CLASSIFICATION:	4P

Notes:

(a) Slight variations for sub-classes.

(b) 52 tons 13 cwt for those with Westinghouse brakes.

version with a boiler pressure of 200lb.

The D49s were named after counties in which the LNER operated and, because its empire extended from Lossiemouth to North Greenwich, there were still some possibilities left in reserve, even after the naming of thirty-six locomotives. One of the possible names which was not used was Durham, the very county in which the locomotives were built. Despite the fact that the names of the first thirty-three D49s showed consistency by ending in 'shire', there was a vociferous lobby to have Durham commemorated but the LNER's hierarchy simply would not yield to the idea of non-standard nameplates, probably as it was considered insufficient to justify yet another sub-class. The Durham fan club was far from amused when the final three of 1929, Nos 2758/59/60, appeared wearing the distinctly

non-'shire' names of *Northumberland, Cumberland* and *Westmorland*.

The Scottish contingent of D49s was divided between the Edinburgh sheds of Haymarket and St Margarets, Eastfield in Glasgow, Perth and Dundee. Apart from transfers to Carlisle and Thornton, their homes hardly changed for their entire working lives. The Scottish routes did not, in the main, enable the D49s to show what they could really do in the speed stakes but, when it came to haulage power, there were instances of D49s hauling 435-ton trains unaided from Newcastle to Edinburgh. That jolly little challenge was not presented to the D49s very regularly but, on the Aberdeen route, they were frequently put in charge of 390-ton trains.

The English-based D49s of 1927–29 were initially divided between the former North Eastern sheds at

TABLE 2.2: THE D49 CLASS SHIRE LOCOMOTIVES

FIRST NO.	1946 NO.	BR NO.	NAME	BUILT	SHED ALLOCATIONS		WDN.
					1/1/48	1/4/58	
234	2700	62700	*Yorkshire*	10/27	Hull B.G.	Hull B.G.	10/58
251	2701	62701	*Derbyshire*	11/27	Hull B.G.	Hull B.G.	9/59
253	2702	62702	*Oxfordshire*	11/27	St. Margarets	York	11/58
256	2703	62703	*Hertfordshire*	12/27	Hull B.G.	Hull B.G.	6/58
264	2704	62704	*Stirlingshire*	12/27	Thornton Junction	Thornton Junction	8/58
265	2705	62705	*Lanarkshire*	12/27	Haymarket	Haymarket	11/59
266	2706	62706	*Forfarshire*	12/27	Haymarket		2/58
236	2707	62707	*Lancashire*	1/28	Hull B.G.	Hull B.G.	10/59
270	2708	62708	*Argyllshire*	1/28	Thornton Junction	Thornton Junction	5/59
277	2709	62709	*Berwickshire*	1/28	Haymarket	Haymarket	1/60
245	2710	62710	*Lincolnshire*	2/28	Hull B.G.	Hull B.G.	10/60
281	2711	62711	*Dumbartonshire*	2/28	Haymarket	St. Margarets	5/61
246	2712	62712	*Morayshire*	2/28	Haymarket	Thornton Junction	7/61
249	2713	62713	*Aberdeenshire*	2/28	Dundee		9/57
250	2714	62714	*Perthshire*	3/28	Perth South	Stirling South	8/59
306	2715	62715	*Roxburghshire*	3/28	St. Margarets	St. Margarets	6/59
307	2716	62716	*Kincardineshire*	3/28	Thornton Junction	Thornton Junction	4/61
309	2717	62717	*Banffshire*	3/28	Thornton Junction	Hull B.G.	1/61
310	2718	62718	*Kinross-shire*	5/28	Dundee	St. Margarets	4/61
311	2719	62719	*Peebles-shire*	5/28	Haymarket	Haymarket	1/60
318	2720	62720	*Cambridgeshire*	5/28	Hull B.G.	Hull B.G.	10/59
320	2721	62721	*Warwickshire*	5/28	Haymarket	St. Margarets	8/58
322	2722	62722	*Huntingdonshire*	7/28	Hull B.G.	Hull B.G.	10/59
327	2723	62723	*Nottinghamshire*	7/28	Hull B.G.	Hull B.G.	1/61

The first of Gresley's N2 0-6-2Ts was built for the Great Northern Railway in December 1920 but construction of similar locomotives continued well into LNER days. However, the only preserved example is a GNR original. In February 1921, the North British Locomotive Co delivered GNR No 1744 which became LNER No 4744 in 1924; under the LNER renumbering scheme of 1946, it became No 9523 and, at Nationalisation, No 69523 was applied. The locomotive was considered past its sell-by date in September 1962, by which time it had outlived almost all of its classmates, and was bought by the Gresley Society and subsequently sent to live on the Keighley & Worth Valley Railway. It now resides at the Great Central Railway centre at Loughborough and this is where it was photographed in July 1985.

Photo: P. Chancellor

The sole N2 in preservation is, at the time of writing, undergoing an overhaul at the Great Central Society's workshop at Loughborough. This photograph shows the locomotive wearing its British Railways livery, complete with No 69523, and this can be compared to the other picture of the same locomotive.

Photo: Peter Herring

Of the seventy-six D49 4-4-0s which were built, only one is preserved. The last of the class to be retired was No 62712 *Morayshire* which was withdrawn from service on 3 July 1961 but it continued to be used as a stationary boiler for an Edinburgh laundry until January 1962. Although Nigel Gresley was as keen as the next man to see his washing done well, he would probably have been quite unamused by his locomotive's undignified task. When its laundry days were over, the D49 was stored and, two and a half years later, it was purchased for preservation and restored almost to its original condition, complete with No 246. There were a few minor departures from the description of 'original condition' although the only blatant one was the pairing with a Great Central-type tender but, admittedly, *Morayshire* had worked with that same tender since June 1941. This picture shows an immaculate No 246 towing equally spotless ex-Caledonian Railway 0-4-4T No 419 at Darlington in 1975.

Photo: E. H. Sawford

There is only one non-streamlined Gresley Pacific in preservation but, arguably, it is one of the most famous preserved locomotives of them all. Soon after A3 No 60103 *Flying Scotsman* had been withdrawn in January 1963, it was purchased by Mr. Alan Pegler but restoration to full original condition was totally impossible. The locomotive had started life as an A1 in February 1923 and been rebuilt as an A3 in January 1947 but, as the last new A1 boiler had been constructed as far back as 1925, Mr. Pegler quite happily settled for a compromise. This in no way detracted from the overwhelming appeal of the restored locomotive, particularly as other details had received a surfeit of attention. For example, the trough deflectors of 1961 were removed, the double chimney of 1959 was replaced by a single chimney and, furthermore, the tender with which it had run since 1938 was replaced by an original 1928 corridor tender. As No 4472, it returned to the open road in March 1963 and this picture shows it at Cambridge on 1 October the following year.

Photo: E. H. Sawford

TABLE 2.2: CONTINUED

FIRST NO.	1946 NO.	BR NO.	NAME	BUILT	SHED ALLOCATIONS		WDN.
					1/1/48	1/4/58	
335	2724	62724	*Bedfordshire*	8/28	Hull B.G.		12/57
329	2725	62725	*Inverness-shire*	8/28	Perth South	Stirling South	11/58
352	2726	62726	**Leicestershire*	3/29	Starbeck		12/57
336	2727	62727	**Buckinghamshire*	6/29	Hull B.G.	Starbeck	1/61
2753	2728	62728	*Cheshire*	2/29	Dundee	Thornton Junction	10/59
2754	2729	62729	*Rutlandshire*	4/29	Thornton Junction	Thornton Junction	5/61
2755	2730	62730	*Berkshire*	3/29	Carlisle	York	12/58
2756	2731	62731	*Selkirkshire*	3/29	Carlisle	Selby	4/59
2757	2732	62732	*Dumfries-shire*	3/29	Carlisle	Carlisle	11/58
2758	2733	62733	*Northumberland*	3/29	Haymarket	Thornton Junction	4/61
2759	2734	62734	*Cumberland*	5/29	Carlisle	Carlisle	3/61
2760	2735	62735	*Westmorland*	6/29	Carlisle	Scarborough	8/58

* Locomotives renamed in 1932:

No 352 became *The Meynell*; No 336 became *The Quorn*.

Sub-Classifications

Nos 318/320/322/327/335/329 (later BR Nos 62720–25 built as D49/3 and rebuilt as D49/1 in 1938.

Nos 352/336 (later BR Nos 62726/27) built as D49/2.

The nameplate *The Oakley* was worn by D49/2 No 62769 which had started life in December 1934 as LNER No 366. Each of the Hunt locomotives was adorned with the figure of a fox but why the creatures seemed to be smiling must remain a mystery. For the record, The Oakley was based in the Bedford/St. Neots area.

Photo: Rail Archive Stephenson

York and Neville Hill in Leeds. Those nice people at Neville Hill lent No 245 *Lincolnshire* to Kings Cross shed for seven months in 1928/29 and, during its spell in London, No 245 regularly worked the West Riding Pullman non-stop to its former home town of Leeds. The class leader, Leeds-based No 234 *Yorkshire*, was not to be outdone by its Scottish counterparts' haulage feats northwards from Newcastle; it was known to have hauled a 480-ton train over the 44 miles from Darlington to York in under forty-nine minutes. Unsurprisingly, the D49s had to be worked very hard to perform such feats and, for the fastest schedules, the insurance of double-heading was usually undertaken. This did little to endear the class to footplate crews, as the rear position on a double-headed turn inevitably resulted in the crews receiving liberal helpings of smoke and coal dust from the leading engine.

In the operating department, it was considered that, as long as the raised voices of grime-encrusted footplatemen could be ignored, the D49s were useful engines and so, between April 1932 and

February 1935, a further forty were constructed. The new arrivals all had Lentz rotary gear and were, consequently, designated D49/2. Although a few suitable county names remained unused including, controversially, Durham, the new D49s were all named after hunts and, in order to preserve the demarcation lines of the precious sub-classification system, the two earlier locomotives which had been designated D49/2 had their names changed in 1932. No 336 *Buckinghamshire* and No 352 *Leicestershire* were renamed *The Quorn* and *The Meynell* respectively.

When it came to naming the new D49s after hunts, the East Midlands and rural parts of Yorkshire yielded a number of names which had geographical relevance for LNER locomotives, the LNER had to cast its net well beyond its own area in order to come up with enough names. Among the locomotives which had, by necessity, to display non-local names were No 258 *The Cattistock* and No 361 *The Garth*, the respective location of those hunts being north Dorset and the Hampshire/Berkshire frontier.

The Lentz rotary valve gear can be clearly seen on this neglected D49. No 62737 *The York and Ainsty* was actually the second of the official Hunt series to have been built although two older Shires which carried the Lentz gear were subsequently renamed after hunts. The shedplate on the engine is that of Botanic Gardens but the photograph was taken outside Darlington works. The picture is undated but, as No 62737 was one of the first D49s to be retired (January 1958), it is not impossible that its grubby condition reflects the impending fate of the scrapyard.

Photo: E. H. Sawford

TABLE 2.3: THE D49/2 CLASS HUNT LOCOMOTIVES

FIRST NO.	1946 NO.	BR NO.	NAME	BUILT	SHED ALLOCATIONS		WDN.
					1/1/48	1/4/58	
201	2736	62736	*The Bramham Moor*	4/32	Gateshead	Starbeck	6/58
211	2737	62737	*The York and Ainsty*	5/32	Hull B.G.		1/58
220	2738	62738	*The Zetland*	5/32	Gateshead	Starbeck	9/59
232	2739	62739	*The Badsworth*	5/32	Gateshead	Scarborough	10/60
235	2740	62740	*The Bedale*	6/32	York	Leeds N.H.	8/60
247	2741	62741	*The Blankney*	7/32	Hull B.G.	Hull B.G.	10/58
255	2742	62742	*The Braes of Derwent*	8/32	Gateshead	Leeds N.H.	11/58
269	2743	62743	*The Cleveland*	8/32	Hull B.G.	Haymarket	5/60
273	2744	62744	*The Holderness*	10/32	Hull B.G.	Thornton Junction	12/60
282	2745	62745	*The Hurworth*	10/32	Gateshead	York	3/59
283	2746	62746	*The Middleton*	8/33	Leeds N.H.	Starbeck	5/58
288	2747	62747	*The Percy*	8/33	Gateshead	York	3/61
292	2748	62748	*The Southwold*	8/33	Leeds N.H.		12/57
297	2749	62749	*The Cottesmore*	8/33	Gateshead	Leeds N.H.	7/58
298	2750	62750	*The Pytchley*	9/33	Gateshead	Hull B.G.	11/58
205	2751	62751	*The Albrighton*	7/34	York	Scarborough	3/59
214	2752	62752	*The Atherstone*	7/34	Starbeck	Starbeck	7/58
217	2753	62753	*The Belvoir*	7/34	Starbeck	Starbeck	9/59
222	2754	62754	*The Berkeley*	7/34	Hull B.G.	Hull B.G.	11/58
226	2755	62755	*The Bilsdale*	7/34	York	Selby	11/58
230	2756	62756	*The Brocklesby*	8/34	Leeds N.H.		3/58
238	2757	62757	*The Burton*	8/34	Hull B.G.		12/57
258	2758	62758	*The Cattistock*	8/34	Leeds N.H.		12/57
274	2759	62759	*The Craven*	8/34	York	Starbeck	1/61
279	2760	62760	*The Cotswold*	9/34	York	York	10/59
353	2761	62761	*The Derwent*	9/34	York		12/57
357	2762	62762	*The Fernie*	9/34	Starbeck	Starbeck	10/60
359	2763	62763	*The Fitzwilliam*	9/34	York	Starbeck	1/61
361	2764	62764	*The Garth*	10/34	Gateshead	Leeds N.H.	11/58
362	2765	62765	*The Goathland*	10/34	Leeds N.H.	Starbeck	1/61
363	2766	62766	*The Grafton*	11/34	Gateshead	Hull B.G.	9/58
364	2767	62767	*The Grove*	11/34	Hull B.G.	Hull B.G.	10/58
365	*2768	62768	*The Morpeth*	12/34	Starbeck		11/52
366	2769	62769	*The Oakley*	12/34	Leeds N.H.	Scarborough	9/58
368	2770	62770	*The Puckeridge*	12/34	Leeds N.H.	Scarborough	9/59
370	2771	62771	*The Rufford*	1/35	Gateshead	York	10/58
374	2772	62772	*The Sinnington*	1/35	Leeds N.H.	Selby	9/58
375	2773	62773	*The South Durham*	1/35	Starbeck	Leeds N.H.	8/58
376	2774	62774	*The Staintondale*	2/35	Leeds N.H.	Leeds N.H.	11/58
377	2775	62775	*The Tynedale*	2/35	Leeds N.H.	Selby	12/58

* No 365 (later BR No 62768) rebuilt as D class in 1942.

Table 129.

DONCASTER AND HULL.

This LNER timetable for the summer of 1938 shows the services from Doncaster to Hull. The D49 4-4-0s were used extensively on these services at that time.

The new D49s were all allocated south of the border and, apart from small contingents at Gateshead and Heaton, were accommodated in Yorkshire. There was a knock-on effect in that, when the new machines arrived at the Leeds and York depots, older D49s were transferred elsewhere. The Yorkshire-based D49s had their share of express duties northwards to Newcastle and southwards to Grantham, but the major proportion of their work was on runs of between fifty and one hundred miles with trains of seven to ten coaches. Among their regular semi-fast cross-country workings were the runs between Hull and Sheffield, Hull and York, Leeds and Scarborough and Newcastle and Carlisle.

In Scotland, the D49s tended to be used on duties which had previously been the strongholds of 4-4-2s, but the 4-4-0s had their loadings limited to, on average, twenty to forty tons less than that of the 4-4-2s. The workings undertaken most regularly by the Edinburgh-based D49s were over the Waverley route to Carlisle and on express turns to Dundee, Glasgow and Perth. The D49s which lived at Dundee tended to operate northwards to Aberdeen but this often required the grimy practice of double-heading.

The D49s were, generally, robust engines but they did have one inherent fault. In common with some of Gresley's other three-cylinder engines, the middle big-end bearing had a tendency to run hot and, in order to provide an early-warning system for the driver, Gresley designed a fluid container which could be fitted inside the crank axle. When things started getting a little too warm, the heat caused the fluid to give off a pungent gas and, even on runs over the Waverley route during a winter gale, the gas was strong enough to be detected in the cab. The ingenious device might have prevented breakdowns and costly repairs but the crews' references to 'Nigel's Stink Bombs' spoke volumes for the nature of the aroma. Even when the fragrance was changed from violets to aniseed, the crews were not impressed.

The usual story throughout the railways of Britain was that express locomotives would only retain their top spots until the arrival of newer, more powerful machines. In England the D49s had stiff competition in the form of Gresley's 4-6-2s and, later, V2s and B1s but, in Scotland, 4-4-0s were more suited to some routes than 4-6-2s and so

the Scottish D49s maintained a high profile for slightly longer than their English counterparts. During the war years, however, the Scottish D49s started to see occasional use on freight duties and this signalled the beginning of the end for their previously lofty status. Wartime traffic in England put a considerable strain on the Yorkshire-based D49s as they struggled to cope with some of the loadings and, in common with the Scottish contingent, they did not recover their express status after the war.

During the war years, No 365 *The Morpeth* underwent a transformation. In August 1942, Gresley's successor, Edward Thompson, rebuilt the locomotive with just two cylinders instead of the original three but the experiment was not successful. The best that can be said for No 365 was that it gave the excuse for yet another sub-classification which, unofficially, was D49/4 but was listed on the books simply as D. Whether D49/4 or just plain old D, *The Morpeth* was withdrawn in November 1952, by which time it carried No 62768. The British Railways renumbering of the class had provided Nos 62700–75.

Surprisingly, withdrawal of the unadulterated D49s did not commence until 1957 but the task was completed when No 62712 (originally LNER No 246) *Morayshire* was retired from Dalry Road on 3 July 1961. Instead of being despatched to the scrapyard, *Morayshire* was put into storage and was purchased privately in 1964. The following year, it was treated to LNER green livery at Inverurie works and now lives on the Bo'ness and Kinniel Railway in West Lothian as the sole representative of the first express passenger class which Nigel Gresley designed for the LNER.

Although it was October 1927 when the first D49 was completed, the drawings for the class had been prepared in February 1926. Between those dates, Gresley fitted a new boiler to one of his 4-6-2s, but this seemingly unexciting event was to prove a significant step towards the eventual evolution of his masterpieces. Early in 1927, Gresley expressed dissatisfaction with the E-type superheaters which had been developed with the 4-6-2s and 2-8-2s in mind but the alternative, the Robinson superheater, was considered to require upgrading if it was to be used on the largest locomotives. The other option

TABLE 2.4: YEARLY TOTALS OF D49 CLASS LOCOMOTIVES

Totals taken at 31 December of each year.

1927	7	1932	46	1937	76	1942	75	1947	75	1952	75	1957	68
1928	26	1933	51	1938	76	1943	75	1948	75	1953	75	1958	38
1929	36	1934	71	1939	76	1944	75	1949	75	1954	75	1959	22
1930	36	1935	76	1940	76	1945	75	1950	75	1955	75	1960	14
1931	36	1936	76	1941	76	1946	75	1951	75	1956	75	1961	0

was to look for a new superheater design, and the one which impressed Gresley most was that used by the 4-6-2s of the German State Railways. In March 1927, Gresley ordered the construction of five new boilers which had not only superheaters similar to the German devices, but also an increase in pressure to 220lb. In July 1927, 4-6-2 No 4480 *Enterprise* was fitted with the first of the new boilers and was consequently reclassified as an A3.

The manner in which Gresley had been reminded that there was scope for improvement of the 4-6-2s had been somewhat embarrassing. Although the A1 4-6-2s had performed their duties well, comparison trials had been conducted against two of the Great Western's Castle class 4-6-0s in 1925 and, much to

the LNER's horror, the Swindon engines had proved superior to their Doncaster counterparts in almost every department. Gresley had not been amused. This had prompted a rethink on the LNER regarding the design of future 4-6-2s, particularly in view of the fact that there was a degree of optimism that the Waverley route could, sooner or later, be opened up permanently to the larger engines.

When No 4480 was fitted with its larger high-pressure boiler as the prototype A3, it was allowed to retain its 6ft 8in driving wheels and 20in × 26in cylinders; consequently, its tractive effort showed an impressive increase from 29,835lb to 36,465lb. When No 2544 *Lemberg* was being fitted with a

Of Gresley's A1s, Nos 2568–82 were earmarked for the North Eastern Area and so they were fitted with Westinghouse brakes for the engine, tender and train. This picture of No 2578 *Bayardo* was taken at Doncaster in September 1932 and although the locomotive had, by then, been rebuilt as an A3, the Westinghouse donkey pump still holds pride of place. The locomotive eventually became BR No 60079 and was retired in September 1961.

Photo: Rail Archive Stephenson

In July 1927, A1 class No 4480 *Enterprise* was fitted with a new boiler and thereby became the prototype A3; early in 1928, three other A1s were reboilered as A3s and one of these was four year-old No 2578 *Bayardo*. Here, No 2578 is seen south of Thirsk in 1932 but, considering the remarkable short rake of carriages, one could be forgiven for thinking that the LNER had adopted the old Midland Railway maxim of 'keep 'em light'. I am reliably informed that the American pattern outer semaphore signals in the picture were removed in 1933.

Photo: E. R. Morten

similar boiler late in 1927, it was decided to line its cylinders down to 18¼in × 24in so that its tractive effort would be within a few hundred pounds of the unmodified A1s, thereby enabling fairer comparisons between the larger- and smaller-boilered 4-6-2s. During the first part of 1928, the lucky recipients of the three other experimental boilers were No 2573 *Harvester*, No 2578 *Bayardo* and No 2580 *Shotover* all of which retained 20in diameter cylinders.

For the tests in February 1928, the smaller-cylindered A3, No 2544, was pitted against un-rebuilt A1 No 4473 *Solario* on express duties between Kings Cross and Doncaster, while No 2580 performed solo trials on the Waverley route, following in the footsteps of A1 No 2563 *William Whitelaw* which had warily traversed the line with a 'fact finding' train a year earlier. The outcome of both sets of tests showed the A3s to be superior to

the A1s but little more so than would have been anticipated with the larger boilers. Nevertheless, an improvement was an improvement and so the building of the old A1 boilers was discontinued.

Gresley was satisfied that the idea behind the A3s was worth pursuing and, between August 1928 and April 1929, ten new A3s were constructed at Doncaster and were numbered 2743–52; eight more brand spanking new A3s were completed between February and July 1930 and were given Nos 2595–99 and 2795–97. The new A3s incorporated several minor differences to the rebuilt locomotives and, from the crews' point of view, the most significant was the use of left-hand drive. The LNER's classification staff were delighted with the new arrivals, as the use of 19in × 26in cylinders enabled another sub-class to be entered into the books. The four reboilered locomotives with 20in diameter cylinders were designated A3/1, the

TABLE 2.5: ORIGINAL DIMENSIONS OF A3 CLASS 4-6-2s.

BUILT:	Doncaster 1928–35
WEIGHTS FULL (locomotive):	96 tons 5 cwt
(tender)	57 tons 18 cwt (a)
TENDER CAPACITY (water):	5,000 gallons
(coal):	8 tons 0 cwt (a)
WHEELBASE (locomotive):	35ft 9in
(tender):	16ft 0in
WHEEL DIAMETERS (bogie):	3ft 2in
(coupled):	6ft 8in
(trailing):	3ft 8in
CYLINDERS:	(3) 19in × 26in (b)
BOILER PRESSURE:	220 lb
HEATING SURFACES (tubes):	1398.8 sq ft
(flues):	1122.8 sq ft
(firebox):	215.0 sq ft
(superheater):	706.0 sq ft
GRATE AREA:	41.25 sq ft
TRACTIVE EFFORT @ 85% boiler pressure:	32,909 lb
LNER ROUTE AVAILABILITY:	9
BR POWER CLASSIFICATION:	7P

Notes:

(a) Corridor tenders, which were fitted to four of the locos at first, weighed 62 tons 8 cwt and had a coal capacity of 9 tons. Six locos had Great Northern-style tenders when new and their weights were 56 tons 6 cwt.

(b) Cylinder sizes of the original rebuilt A1s varied. See text.

rebuild with 18¼in cylinders had sub-class A3/2 all to its very own while the new engines with 19in diameter cylinders became A3/3. Most upsettingly for the classifiers, the five original reboilered engines were all fitted with 19in cylinders in the early 1930s and, therefore, joined the A3/3 sub-class. Despite the A3/1 and A3/2 sub-classes being devoid of any members after August 1934, they were still officially recognised and meticulously documented until December 1938.

The first three A3s which appeared in 1928, Nos 2743–45, were paired with new corridor tenders. Ten such tenders were actually constructed, but seven were completed somewhat faster than the locomotives for which they were intended and so they were fitted, instead, to other 4-6-2s. The LNER was the first railway company in the world to use corridor tenders but the idea for them had come about out of necessity.

For the summer of 1927, the LNER had introduced non-stop workings between Kings Cross and Newcastle, and these had proved so successful that the dream of non-stop services between Kings Cross and Edinburgh seemed as if it could be turned into a reality. Neither Gresley nor the LNER had doubts about the abilities of the A1s and A3s to handle non-stop services to Scotland, but such a duty was considered too risky for a single footplate crew. Although it would not have been impossible to accommodate a relief crew on the footplate, it was considered preferable for the off-duty crew members to spend their rest time on the train instead of in the cab. One route which the crews could use to get from the footplate to the train was via the top of the tender but, although that became popular, in later days, with John Wayne and Clint Eastwood, the LNER conceded that it would not be the safest of practices. The alternative was to

The last of the A1s was built in December 1924 and, just two months previously, LNER No 2572 *Galopin* had emerged from Doncaster works. In June 1941, it was rebuilt as an A3 and, at Nationalisation, became No 60076. Here it is seen at its home shed of Darlington on 7 July 1956 alongside Gresley 2-6-2T No 67618. At the time the picture was taken, the 2-6-2T was still running as a V1; rebuilding as a V3 took place in 1958.

Photo: E. H. Sawford

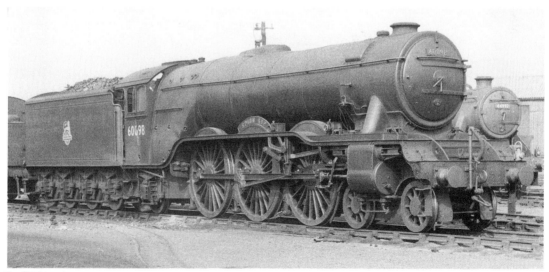

No 60098 *Spion Kop* was one of the early purpose-built A3s. It was delivered in April 1929 and was to live until October 1963. This picture was taken at Perth on 25 August 1955.

Photo: E. H. Sawford

construct a passageway through the tender and use conventional corridor connections to link the rear of the tender to the first carriage.

The corridor tenders came into their own when the non-stop service to Edinburgh, the Flying Scotsman, was inaugurated on 1 May 1928. The first Flying Scotsman left Kings Cross at 10.00 am hauled, most appropriately, by A1 No 4472 *Flying Scotsman* while, at exactly the same time, the southbound train left Edinburgh in the charge of A3 No 2580 *Shotover*. The run was all of 392 ¾ miles and this stole the distinction of Britain's longest non-stop service from the Great Western's Cornish Riviera Express. The loadings of the Flying Scotsman were in excess of four hundred tons but the A1s and A3s were well used to trains of six hundred tons on regular main-line duties.

The LNER would dearly have loved to establish speed records on the service and the locomotives would, no doubt, have been able to oblige their masters, but a long-standing agreement restricted the speeds of Anglo-Scottish expresses. Much to its chagrin, the LNER had to allow eight and a quarter hours for the run and, although the restrictions were

to be eased in 1932, it was not until 1937 that a six-hour schedule could be legitimately introduced. During the summer of 1928, a total of 125 non-stop workings was made and only one arrived late; this was due to a blockage on the line at Chatshill in Northumberland. On one other occasion, an engine failure on a London-bound Flying Scotsman meant a change of locomotive at Grantham, but even this wasn't enough to prevent arrival at Kings Cross on time.

Nine more A3s, Nos 2500–08, were delivered between July 1934 and February 1935 to bring the class total to thirty-two. When the A3's predecessors, the A1s, had been constructed, only four of the locomotives (Nos 2555/63, 4470/72) had been named from new and only one other (No 4471) had been named since construction. Early in 1925, the LNER's directors had decided that such high-profile locomotives should not remain anonymous and, displaying their enthusiasm for the turf, they had decided to use the names of champion racehorses. In the days before the 'ITV Seven' introduced cryptic horse names to the great British public, locomotives bearing the titles *Pretty Polly* and *Merry*

The very last A3 to be built was No 2508 *Brown Jack* which emerged from Doncaster works in February 1935. Here, it is in charge of the southbound Flying Scotsman near Potters Bar on 8 August 1936. The locomotive eventually became BR No 60043 and survived until May 1964 which was longer than many of its classmates.

Photo: LCGB/Ken Nunn Collection

TABLE 2.6: THE A3 CLASS 4-6-2s.

N.B. This list covers only the locomotives which were built as A3s and *not* the rebuilt A1s. The list uses numerical order and *not* the order of construction.

FIRST NO.	1946 NO.	BR NO.	NAME	BUILT	SHED ALLOCATIONS		WDN.
					1/1/48	1/1/60	
2500	35	60035	*Windsor Lad*	7/34	Haymarket	Haymarket	9/61
2501	36	60036	*Colombo*	7/34	Gateshead	Leeds N.H.	11/64
2502	37	60037	*Hyperion*	7/34	Haymarket	Haymarket	12/63
2503	38	60038	*Firdaussi*	8/34	Gateshead	Gateshead	11/63
2504	39	60039	*Sandwich*	9/34	Grantham	Kings Cross	3/63
2505	40	60040	*Cameronian*	10/34	Gateshead	Darlington	7/64
2506	41	60041	*Salmon Trout*	12/34	Haymarket	Haymarket	12/65
2507	42	60042	*Singapore*	12/34	Gateshead	Gateshead	7/64
2508	43	60043	*Brown Jack*	2/35	Haymarket	Haymarket	5/64
2595	84	60084	*Trigo*	2/30	Gateshead	Leeds N.H.	11/64
2596	85	60085	*Manna*	2/30	Heaton	Heaton	10/64
2597	86	60086	*Gainsborough*	4/30	Gateshead	Leeds N.H.	11/63
2598	87	60087	*Blenheim*	6/30	Haymarket	Haymarket	10/63
2599	88	60088	*Book Law*	7/30	Heaton	Heaton	10/63
2743	89	60089	*Felstead*	8/28	Kings Cross	Haymarket	10/63
2744	90	60090	*Grand Parade*	8/28	New England	Haymarket	10/63
2745	91	60091	*Captain Cuttle*	9/28	Carlisle	Gateshead	10/64
2746	92	60092	*Fairway*	11/28	Heaton	Heaton	10/64
2747	93	60093	*Coronach*	12/28	Carlisle	Carlisle	4/62
2748	94	60094	*Colorado*	12/28	Haymarket	Haymarket	2/64
2749	95	60095	*Flamingo*	2/29	Carlisle	Carlisle	4/61
2750	96	60096	*Papyrus*	3/29	Kings Cross	Haymarket	9/63
2751	97	60097	*Humorist*	4/29	Kings Cross	Haymarket	8/63
2752	98	60098	*Spion Kop*	4/29	Kings Cross	Haymarket	10/63
2795	99	60099	*Call Boy*	4/30	Haymarket	Haymarket	10/63
2796	100	60100	*Spearmint*	5/30	Haymarket	Haymarket	6/65
2797	101	60101	*Cicero*	6/30	Haymarket	Haymarket	4/63

Hampton had been considered, surprisingly, safe to work in close proximity. The A3s of 1934/35 were, like their predecessors of 1928–30, named from new and the racehorse theme was continued.

The trials of 1928 had shown that, while the A3s were superior to the A1s, the difference had been far from earth-shattering and so the A3s tended to share the LNER's crack turns with the A1s rather than replace them. Predictably, it had been the main-line sheds which had received most of the A3s from new but, in 1928/29, two had been sent

new to Carlisle to join another which had been transferred there in 1927. The Waverley route had, by then, been opened up for 4-6-2s and their ability to handle loadings up to an official limit of 400 tons over the troublesome line had enabled an extensive improvement of services and a partial elimination of double heading. Elsewhere in Scotland, another problem route had been the steeply graded line between Edinburgh and Aberdeen but, although an A3 had made tentative test runs as far as Montrose in 1928, it was not until the LMS had strengthened the line northwards from Montrose in 1930 that the 4-6-2s could be used regularly on the route.

In 1935, the first of Gresley's famous A4s appeared, but the A3s were not ousted from turns on the Flying Scotsman until 1937. The introduction of the A4s did not, however, mean that the A3s were relegated en masse to the second division as there were plenty of express duties available for them. In 1937, an express service was introduced between Glasgow and Leeds, and two A3s were transferred to Eastfield shed to help out on these turns. At some Scottish depots, the job of turning the A3s was not the simplest of operations due to the absence of suitable turntables. At Haymarket shed in Edinburgh, a seventy-foot turntable had been installed in 1931 but St. Margarets had to wait until 1942 for one of its own. Fortunately, there was a choice of two triangles in Edinburgh, one to the east of Waverley station and the other to the west of Haymarket shed, which were available for turning. In Perth, the former North British depot possessed a turntable of only fifty-two feet and, as the shed yard was so cramped, there was no room

for a larger one to be installed, but the LMS shed along the line was kind enough to let the LNER 4-6-2s use its full-size turntable.

The order for the A4s was placed in March 1935 and, as if to take the limelight away from what was anticipated to be Gresley's most far-reaching design, A3 No 2750 *Papyrus* staked its own claim to a place in the history books. On 5 March 1935, a test run was made prior to the introduction of the Silver Jubilee train and the target was a four hour timing for the 269 miles between Kings Cross and Newcastle. The northbound run was made three minutes under schedule which, considering that there was a delay because of a derailed goods train, was highly impressive but the fireworks were kept in hand for the homeward run. It wasn't just that the arrival at Kings Cross was eight minutes early but a speed in excess of 100 mph was sustained for over twelve miles between Corby and Tallington, and this included an average of over 105 mph for the four miles between Little Bytham and Essendine. Over the entire length of the journey from Newcastle to Kings Cross, the average speed was 69.4 mph. Had

Here's a combination and a half for Gresley enthusiasts. A3 No 2507 *Singapore* and one of the Silver series A4s are seen passing Belle Isle signal box while backing down in tandem to Kings Cross. The date is 8 June 1937.

Photo: LCGB/Ken Nunn Collection

it not been for the arrival of the A4s later that year, the achievement of No 2750 would have remained in the forefront of locomotive running for a more respectable period.

The construction of A1 boilers had ceased in 1925 and so it had seemed likely that the conversion of A1s to A3s would have been undertaken, slowly but surely. However, it was November 1939 before the next conversion took place and the practice was not introduced wholesale until June 1941, by which time Nigel Gresley had been succeeded by Edward Thompson. The last to be converted was BR No 60068 (ex-LNER No 2567) *Sir Visto* which was treated in December 1948 although, from 1945, *Sir Visto* and the eleven other unrebuilt A1s which remained had all been reclassified as A10s. The only A1 which did not resurface as an A3 was the original Gresley 4-6-2, No 4470 (ex-GN No 1470) *Great Northern*, which was rebuilt by Thompson as the first of a new A1 class in 1945.

TABLE 2.7: GRESLEY A1 PACIFICS REBUILT AS A3s.

N.B. In April 1945, the remaining unrebuilt A1s were reclassified A10.

BR NO.	NAME	FIRST NO.	DATE RBLT.	BR NO.	NAME	1ST NO.	DATE RBLT.
60044	*Melton*	2543	9/47	60070	*Gladiateur*	2569	1/47
60045	*Lemberg*	2544	12/27	60071	*Tranquil*	2570	10/44
60046	*Diamond Jubilee*	2545	8/41	60072	*Sunstar*	2571	7/41
60047	*Donovan*	2546	1/48	60073	*St. Gatien*	2572	11/45
60048	*Doncaster*	2547	5/46	60074	*Harvester*	2573	4/28
60049	*Galtee More*	2548	10/45	60075	*St. Frusquin*	2574	6/42
60050	*Persimmon*	2549	12/43	60076	*Galopin*	2575	6/41
60051	*Blink Bonny*	2550	11/45	60077	*The White Knight*	2576	7/43
60052	*Prince Palatine*	2551	8/41	60078	*Night Hawk*	2577	1/44
60053	*Sansovino*	2552	9/43	60079	*Bayardo*	2578	5/28
60054	*Prince of Wales**	2553	7/43	60080	*Dick Turpin*	2579	11/42
60055	*Woolwinder*	2554	6/42	60081	*Shotover*	2580	2/28
60056	*Centenary*	2555	8/44	60082	*Neil Gow*	2581	1/43
60057	*Ormonde*	2556	1/47	60083	*Sir Hugo*	2582	12/41
60058	*Blair Athol*	2557	12/45	60102	*Sir Frederick Banbury*	4471	10/42
60059	*Tracery*	2558	7/42				
60060	*The Tetrarch*	2559	1/42	60103	*Flying Scotsman*	4472	1/47
60061	*Pretty Polly*	2560	5/44	60104	*Solario*	4473	10/41
60062	*Minoru*	2561	6/44	60105	*Victor Wild*	4474	10/42
60063	*Isinglass*	2562	4/46	60106	*Flying Fox*	4475	3/47
60064	*Tagalie**	2563	11/42	60107	*Royal Lancer*	4476	10/46
60065	*Knight of Thistle**	2564	3/47	60108	*Gay Crusader*	4477	1/43
60066	*Merry Hampton*	2565	12/45	60109	*Hermit*	4478	11/43
60067	*Ladas*	2566	11/39	60110	*Robert the Devil*	4479	8/42
60068	*Sir Visto*	2567	12/48	60111	*Enterprise*	4480	7/27
60069	*Sceptre*	2568	5/42	60112	*St. Simon*	4481	8/46

* Renamed locomotives:
  LNER No 2553 (later BR No 60054) named *Manna* until 12/26.
  LNER No 2563 (later BR No 60064) named *William Whitelaw* until 7/41.
  LNER No 2564 (later BR No 60065) named *Knight of the Thistle* until 12/32.

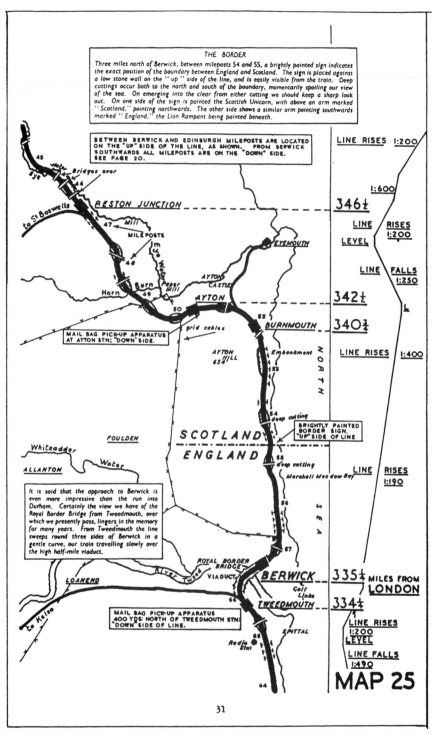

THE BORDER

Three miles north of Berwick, between mileposts 54 and 55, a brightly painted sign indicates the exact position of the boundary between England and Scotland. The sign is placed against a low stone wall on the " up " side of the line, and is easily visible from the train. Deep cuttings occur both to the north and south of the boundary, momentarily spoiling our view of the sea. On emerging into the clear from either cutting we should keep a sharp look out. On one side of the sign is painted the Scottish Unicorn, with above an arm marked " Scotland," pointing northwards. The other side shows a similar arm pointing southwards marked " England," the Lion Rampant being painted beneath.

BETWEEN BERWICK AND EDINBURGH MILEPOSTS ARE LOCATED ON THE "UP" SIDE OF THE LINE, AS SHOWN. FROM BERWICK SOUTHWARDS ALL MILEPOSTS ARE ON THE "DOWN" SIDE. SEE PAGE 20.

LINE RISES 1:200

1:600

346¼

LINE RISES 1:200

LEVEL

LINE FALLS 1:250

342½

340¾

RESTON JUNCTION

EYEMOUTH

AYTON CASTLE

AYTON

BURNMOUTH

MAIL BAG PICK-UP APPARATUS AT AYTON STN: "DOWN" SIDE.

LINE RISES 1:400

AYTON 65¾MILL

Embankment

BRIGHTLY PAINTED BORDER SIGN. "UP" SIDE OF LINE

SCOTLAND
ENGLAND

deep cutting

deep cutting

LINE RISES 1:190

FOULDEN

Whiteadder Water

ALLANTON

Marshall Meadow Bay

It is said that the approach to Berwick is even more impressive than the run into Durham. Certainly the view we have of the Royal Border Bridge from Tweedmouth, over which we presently pass, lingers in the memory for many years. From Tweedmouth the line sweeps round three sides of Berwick in a gentle curve, our train travelling slowly over the high half-mile viaduct.

ROYAL BORDER BRIDGE

VIADUCT

LOANEND

BERWICK

TWEEDMOUTH

Golf Links

335¼ MILES FROM LONDON

334¾

MAIL BAG PICK-UP APPARATUS 400 YDS. NORTH OF TWEEDMOUTH STN: "DOWN" SIDE OF LINE.

SPITTAL

Radio Stat

LINE RISES 1:200
LEVEL

LINE FALLS 1:490

to Kelso

MAP 25

31

For two shillings (10p), the LNER's passengers could buy a booklet entitled 'Mile by Mile' which extolled the delights of the East Coast main line. This page from the 1947 publication describes points of interest on the section which crosses the border into Scotland.

By the time of Nationalisation, the A3 class comprised seventy-eight locomotives and, apart from six which spent short periods in the experimental British Railways livery of ultramarine, all soon had their LNER green liveries replaced by BR's more sensible idea of Brunswick green. Under Edward Thompson's renumbering scheme of 1943, the A3s were due to take Nos 501–78, but only twenty received their new numbers before the revised renumbering of 1946 gave them Nos 35–112. Whereas the 1943 renumbering placed the locomotives in order of seniority, the 1946 scheme followed the order of their original numbers. The first batch, Nos 35–43, had originally been Nos 2500–08 and had been the last of the class to be built while the eleven oldest survivors, originally Nos 4471–81, became the tail-enders with Nos 102–12. Nevertheless, it was the final LNER renumbering which prepared the A3s for State ownership and they became BR Nos 60035–60112.

In the first year of Nationalisation, the A3s regained their duties on the Yorkshire Pullman when that train was reintroduced after having been suspended in 1947 due to a coal shortage. In 1949, nine A3s were transferred to the former Great Central line which had seen its first A1s ten years previously. Six of those transferred were allocated to Leicester but the familiar problem of inadequate turntables reared its head again, as the only seventy-foot turntable available to them in the city was not at the shed but at Leicester Central Station. The A3s put in their share of duties on named trains through Leicester and these included the South Yorkshireman and the Master Cutler. During the 1950s, the A3s retained many of their traditional duties although the A4s and the new A1s tended to monopolise the most prestigious turns, such as the Flying Scotsman and the Capitals Limited. By the late 1950s, however, the A3s were doing sterling work on other named expresses such as the Waverley, the Talisman and the Thames-Clyde.

Between 1958 and 1960, double chimneys became

Trough deflectors appeared only very late in the lives of the A3s and not all members of the class survived long enough to have them fitted. This picture of No 60037 *Hyperion* was taken on 13 July 1962, two months after it received its deflectors but, by then, withdrawals had already made considerable gaps in the list of A3s. No 60037 was to survive until December 1963. The picture was taken at Haymarket shed but, interestingly, the shedplate carried is that of St. Margarets. But never mind; they're both in Edinburgh.

Photo: M. John Stretton

standard for the A3s after experiments in 1956 had proved that the Great Western Railway's idea of modified blastpipes and double chimneys improved steaming. One of the A3s, No 60097 (ex-LNER No 2751) *Humorist* had in fact been fitted with a double chimney in 1937 and, although the experiment had proved inconclusive, the fitment remained in position. One problem which the LNER had found with the double chimney of *Humorist* was that, as the exhaust pressure had been less than that of a single chimney, the smoke had tended not to be raised enough to prevent obstruction of the driver's view. In the 1950s, British Railways found exactly the same problem of impaired visibility with the double chimney A3s and, as it was considered somewhat advantageous for an engine driver to see where he was going, the idea of smoke deflectors was raised again. Small wing-type deflectors, similar to ones which had once been fitted to *Humorist*, were tried on four A3s in 1959 but were found to be ineffective. The following year, the Kings Cross shedmaster came up with the suggestion of German-style trough deflectors and this proved to be the solution. In 1961, authorisation was given for the fitting of trough deflectors to all the A3s except for the four which were working on the Midland Region. Not all of the A3s, however, were to survive long enough to be given their deflectors.

With the introduction of the Deltic diesels in 1961, the future looked grim not just for the A3s but also the other former LNER 4-6-2s. There had been one withdrawal in December 1958 when No 60104 *Solario* had been retired but the real slaughter did not commence until 1961. By June 1963, steam workings were officially banned south of

Peterborough but the A3s continued to make occasional forays into London until November 1964, by which time an A3 was most definitely on the endangered species list. Only three of the class, Nos 60041/52 and 60100, survived to see 1965 and, by then, they were relegated to freight workings from their home depot of St. Margarets in Edinburgh. The appearance of No 60052 *Prince Palatine* on passenger workings late in 1965 did not, however, signal a miraculous reprieve because, when its two chums were retired at the end of the year, it left No 60052 as the only surviving A3. Its status as such was shortlived as it was retired on 22 January 1966.

Just over three years before No 60052 was withdrawn, the most famous member of the class, No 60103 *Flying Scotsman*, performed its last revenue earning duty for British Railways. On 14 January 1963, it hauled the 1.20 pm from Kings Cross to Leeds but, of course, that was not the end for the locomotive. It was bought by Mr. Alan Pegler and restored to its former splendour, complete with its original number 4472, and has since hauled countless special trains both in this country and abroad. One major distinction which was given to the preserved locomotive was custody of the last steamhauled non-stop Flying Scotsman on 1 May 1968, exactly forty years after it had performed the inaugural duty, albeit as an A1.

The A3s were undoubtedly excellent locomotives but pedantic historians have countered by pointing out that they were, in effect, no more than reboilered versions of older machines. Wherever one's railway loyalties lie, there is no disputing that they evolved because of Gresley's constant quest for improvement and willingness to experiment. At

TABLE 2.8: YEARLY TOTALS OF A3 CLASS LOCOMOTIVES (including rebuilds)

Totals taken at 31 December of each year.

1927	2	1933	23	1939	33	1945	66	1951	78	1957	78	1962	59
1928	11	1934	31	1940	33	1946	70	1952	78	1958	78	1963	26
1929	15	1935	32	1941	39	1947	77	1953	78	1959	77	1964	3
1930	23	1936	32	1942	49	1948	78	1954	78	1960	77	1965	1
1931	23	1937	32	1943	56	1949	78	1955	78	1961	71	1966	0
1932	23	1938	32	1944	61	1950	78	1956	78				

The first ten of Gresley's B17 Sandringhams were constructed by the North British Locomotive Co and one of these was LNER No 2805 *Burnham Thorpe*. This engine was later renamed *Lincolnshire Regiment* and, under State ownership, became BR No 61605. Although the B17s were initially intended for work in East Anglia, the scenery in this photograph provides ample evidence that they went walkabouts. The picture was taken at Torside, between Manchester and Sheffield, in 1939 but, as the train is bound for Harwich, at least the locomotive will eventually see less-challenging terrain.

Photo: E. R. Morten

the time that preparatory work was in hand for the first purpose-built A3s, Gresley and his workforce had their hands completely full and this escalated to problem proportions. In 1928, it was realised that the construction of new locomotives for the former Great Eastern section could not be postponed any longer and, although Gresley would have much preferred to tackle a new design himself, he simply did not have the time and space to take on an additional major task. Instead, he approached the North British Locomotive Company in Glasgow and asked them to design and build a new class of locomotive for him.

The end result of the Scottish design was the popular B17 class 4-6-0s, better known as the Sandringhams. The B17s were required as replacements for James Holden's B12s which, by the late 1920s, were struggling hard to cope with the loadings in East Anglia. Gresley had complete con-

fidence in the ability of the North British Locomotive Company, particularly as the firm had done an excellent job in building some of his A1s and O2s, but the specification for the three-cylinder 4-6-0s included a strict limit on axle loading of just seventeen tons. There was no leeway with the axle weight as many of the bridges on the former Great Eastern lines simply could not accept anything greater but, after a couple of false starts, a design was finalised.

Several standard LNER components were incorporated in the design for the B17s and the first of the class, No 2800, was delivered late in 1928. Only ten were constructed by the North British but, between 1930 and 1937, a further fifty-three of these very useful engines were built at Darlington and ten more by Robert Stephenson & Co. The original design cannot legitimately be credited to Gresley, although it has often been suggested that,

B17 Sandringham No 61627 *Aske Hall* was one of the Darlington-built members of the class. Although the B17s were credited to Nigel Gresley, a number of the class were reboilered and subsequently reclassified in the 1940s by Gresley's successors, Edward Thompson and Arthur Peppercorn. This locomotive was photographed in the north bay at Cambridge on 25 February 1952 after its 'Thompsonisation'.

Photo: E. H. Sawford

The final series of Sandringhams was named after football teams and this shows the nameplate of No 61664. It is known that at least two of the 'football' nameplates finished up in the boardrooms of their respective clubs but, as I have the misfortune to follow a struggling Second Division team, I cannot vouch for the wall decorations at Anfield.

Photo: E. H. Sawford

had he been able to look after the project himself, the tight restrictions which had to be applied to the design would have meant that his own versions would have, almost by necessity, been quite similar to the North British ones. The 6ft 8in coupled wheels and the three 17½in × 26in cylinders of the B17s were distinctly Gresley-ish, but the divided drive, with the outside cylinders powering the middle coupled axle and the inside cylinder doing its stuff on the leading coupled axle, could not have been passed off as Nigel's work.

A taste of Gresley's enthusiasm for experimentation had manifested itself with the A3 class 4-6-2s. His disregard for the orthodox had been seen with the two P1 class 2-8-2s in 1925 which had been, if anything, too powerful for their potential to be used to maximum effect. Gresley's next venture into experimentation on a grand scale was in the planning stage even before the first new A3 had been constructed and, when the new creation was unveiled, it kept the railway journalists of the day adequately supplied with publication cheques.

By 1929, the staff at Darlington works knew something different was in the pipeline when they were given the duty of assembling a set of frames which dwarfed anything that had been seen at the workshop before. For those members of the workforce with an inquisitive nature, the order that the project was to be regarded as 'hush hush' was not the most sensible way of stifling their curiosity. The only hints as to the nature of the new contraption were, firstly, that it was big, secondly, that it was a compound and, thirdly, that it didn't seem to have a boiler. While progress was being made on the chassis of the new locomotive, there was no sign at all of the construction of its boiler and the management was giving no clues. In April 1929, the mystery chassis was removed from the works and, when it reappeared six months later, it was fitted with a ready-constructed boiler. On its return to Darlington, the workforce's initial reaction to the addition of the boiler was a highly indignant shout of 'why weren't we given the job?'. This soon changed to 'what the hell is it anyway?'.

The secret boiler for the 'hush hush' locomotive was of the water-tube type and it had been constructed by Messrs. Yarrow & Co. of Glasgow. The idea behind the design of the boiler was that radiant heat would be directly applied to a far greater proportion of the heating surfaces than with a conventional boiler and significant savings in fuel

Gresley's 'Galloping Sausage' was a remarkable looking machine and, understandably, it was in demand for exhibition purposes. Some said that was just as well as the contraption was always more reliable when stationary. Between January 1930 and May 1935, the locomotive appeared at fifteen different exhibitions but, unfortunately, the location in this picture is unknown.

Photo: E. R. Morten

would, therefore, be made. The use of high-pressure water-tube boilers was commonplace on ships and Harold Yarrow, the chairman of the Glasgow company which built the one for the Darlington locomotive, had supplied many to the Royal Navy and was considered to be Britain's leading expert in the field. Nigel Gresley had first approached Yarrow in 1924 to discuss the possibility of using the marine boilers on railway locomotives and, at the end of that same year, a 2-8-0 freight engine had appeared on the Delaware & Hudson Railroad in America complete with a high-pressure water-tube boiler. Gresley was intrigued, particularly as Harold Yarrow had acted as consultant during the design of the boiler.

Gresley's original plans for the revolutionary locomotive incorporated the use of two low-pressure outside cylinders and a high-pressure one inside. After extensive brain-picking sessions with various acknowledged experts including, embarrassingly, the manager of the Midland Railway's works at Derby, Gresley realised that the locomotive would have to have four cylinders and the optimum dimensions seemed to be 12in × 26in for the high-pressure inside ones and 20in × 26in for the low-pressure outside ones. In the final stages of development, Gresley designed a system whereby all four cylinders were operated by just two outside sets of Walschaerts valve gear with the valves for each set of cylinders being able to be operated independently.

Preliminary discussions between Gresley and Yarrow had involved the assumption that the boiler would be for use on a 4-6-2 but, when the initial drawings were made, Yarrow came up with one snag. The steam drum, which had to be fitted at the highest possible point inside the boiler casing, would have to be thirty feet long and, although there would be no problem in constructing such a drum, fitting it to a standard 4-6-2 would present the minor inconvenience of leaving no space on the footplate for the crew. Gresley considered that the proposed locomotive would do better with the ministrations of a driver and fireman and so, in order to accommodate a usable cab, he altered his plans to include longer frames and, consequently, a second trailing axle. This resulted in what was, nominally, a 4-6-4 but the additional trailing axle

On 2/3 May 1931, No 10000 was exhibited at Norwich and this head-on view of the static exhibit shows why the engine's looks created such a stir.

Photo: LCGB/Ken Nunn Collection

was to be mounted on a pony truck and, when the machine was eventually constructed, pernickety purists were quick to point out to those of a less pedantic nature that the locomotive was, technically, a 4-6-2-2.

When the boiler had been completed at Yarrow's works in Glasgow, the chassis of the locomotive was despatched from Darlington so that the two pieces could be joined in wedlock to enable extensive static tests to be carried out. These included wind tunnel tests which were considered essential due to the design of the boiler. As the steam drum had to be placed at the highest point possible inside the boiler casing, there was room for nothing more than the most basic chimney and this presented the problem of smoke not being cleared from the driver's field of vision. The wind tunnel tests determined the shape and angle of the deflectors which would be required to keep the driver's view clear of exhaust smoke. For over half of its considerable length, the boiler seemed to grow out of the running plates of the locomotive and, when this effect was combined with the extensive deflectors at the front, the overall appearance was of an imposingly streamlined unit.

Apart from having Gresley's standard 6ft 8in driving wheels, the locomotive was, unquestionably, unlike anything else that had been seen in Britain before, and the secrecy to which it was treated did more for its advance publicity than any amount of media hype could have done. When the partly-assembled machine was due to be towed from Yarrow's back to Doncaster works, it was realised that it would have to traverse LMS lines in Glasgow and so an LMS loading gauge inspection could not be avoided. Fearful of giving the LMS a glimpse of the engine even before his own staff had seen it, Gresley ordered that the boiler and chassis should be covered in sheeting and disguised as much as possible before being viewed by the gauge inspectors.

Construction of the locomotive was completed soon after its return to Darlington and, when the wraps came off, public reaction was no less than had been expected. The machine carried No 10000 and it was finished, surprisingly, in battleship grey but the austere livery was broken up by wide steel bands around the huge boiler and smokebox. It was

classified as W1 and made its first trial trip on 12 December 1929. Experimental trips and subsequent minor modifications occupied most of the first six months of its life but these were punctuated by the obligatory exhibition duties, the first of which was the official press day at Kings Cross on 8 January 1930.

No 10000 formally started earning its keep on 20 June 1930 but, instead of being based at the premier shed at Kings Cross, it was allocated to Gateshead depot. An eight-wheel corridor tender had been constructed for the locomotive so that it could take turns on the non-stop Flying Scotsman between Edinburgh and London, and its first appearance on this duty was at the head of the southbound train on 31 July 1930.

Later that year, the failure of an Edinburgh-based 4-6-2 at Newcastle resulted in No 10000 being loaned to the stranded Scottish crew as the railway equivalent of a self-drive 'get-you-home' vehicle, albeit with a scheduled train attached. The crew members were most inquisitive to see how the fabled beast would perform, but they found it to be a poor steamer. Part of the loan agreement was that No 10000 should be worked back to Newcastle the following day and, in order to improve the steaming of the engine, the driver resorted to the

TABLE 2.9: ORIGINAL DIMENSIONS OF W1 CLASS 4-6-4

BUILT:	Darlington & Glasgow 1929
WEIGHTS FULL (locomotive):	103 tons 12 cwt
(tender):	62 tons 8 cwt
TENDER CAPACITY (water):	5,000 gallons
(coal):	9 tons
WHEELBASE (locomotive):	40ft 0in
(tender):	16ft 0in
WHEEL DIAMETERS (leading):	3ft 2in
(coupled):	6ft 8in
(trailing):	3ft 2in
CYLINDERS (HP inside):	(2) 12in × 26in
(LP outside):	(2) 20in × 26in
BOILER PRESSURE:	450 lb
HEATING SURFACES (firebox):	919 sq ft
(combustion chamber):	195 sq ft
(tubes):	872 sq ft
GRATE AREA:	34.95 sq ft
TRACTIVE EFFORT @ 85% boiler pressure:	32,000 lb

The water-tube W1 No 10000 might have been a huge draw at exhibitions but this picture shows it doing what it did worst, i.e. working. Here, on one of its brief periods in service, it is entering Darlington with an up express in June 1930.

Photo: LCGB/Ken Nunn Collection

'engineman's friend', the jimmy. A jimmy was little more than an innocent looking metal bar but, when wedged in the appropriate position across the blastpipe, a tremendous improvement in steaming was obtained. The devices were so widely used that they were even advertised in the trade press but, as they invariably caused excessive wear on the front end of a locomotive, the use of them was most definitely contrary to the rule book. By sheer coincidence, one of the passengers on the Newcastle-bound train that day was Nigel Gresley and, curious to see how a 'foreign' crew would come to grips with his brainchild, he invited himself on the footplate for the run. Needless to say, Gresley was most impressed with the locomotive's performance and steaming ability and, as he had fully expected a good run anyway, he saw no reason to delve inside a hot and grimy smokebox to look at the blastpipe. Gresley went away happy and the relieved driver retained his job.

During the end of 1930 and the beginning of 1931, No 10000 was usually employed on express workings from Newcastle to Edinburgh, York or Leeds and its excursions to Kings Cross became less

and less frequent as it was not at all happy in charge of 500-ton trains on such a long run. After its turn on the Junior Scotsman into Kings Cross on 25 July 1931, it made no more appearances in London in its current guise. Throughout 1931 and 1932, the gremlins appeared on No 10000 with infuriating regularity and the staff at Gateshead shed came to believe that they had the locomotive merely on a time-share basis with Darlington works. By 1934, No 10000 had been excused from express duties and was usually earmarked for working a stopping service from Newcastle to Edinburgh and returning with an undignified fish train.

Many further modifications were made to No 10000 during its frequent workshop visits in 1934 and 1935, but these failed to work the necessary miracles. Much to the relief of the staff at Gateshead shed, it was announced in 1935 that the locomotive would be transferred to another depot and the general opinion was that Darlington shed would be the most appropriate as it would be a shorter journey from there to the workshop. The charmed recipients of No 10000 were, instead, the staff at

Neville Hill shed in Leeds but they didn't have to wait too long before dispensing with the contraption. On 21 August that year, it was back in Darlington works where it rested until being taken to Doncaster works on 13 October 1936 to be completely rebuilt.

When No 10000 had been built in 1929, there had been plans to name it *British Enterprise* and the nameplates had actually been cast. By 1935, hindsight had shown that by not having fitted the nameplates, a great deal of embarrassment had been saved. Its time in revenue-earning service between 20 June 1930 and 21 August 1935 had comprised 1,888 days and, of those, 1,105 had been spent in Darlington works. The unpopularity of the locomotive had given rise to several alternative suggestions for names instead of *British Enterprise*, some of which are printable. At Gateshead, No 10000 was referred to as the 'Galloping Sausage' while, at Leeds, it was called 'Citrus' as it was considered something of a lemon.

Sadly, Nigel Gresley's bold experiment had not paid off but nobody could point a finger at the precise cause of the 4-8-4's lack of success. Difficulty in maintenance, which was not improved by the compound cylinder arrangement, had certainly contributed to the time the locomotive had spent out of service but there seemed to be no overriding reason for its lack of sparkle. Nevertheless, the harsh lessons about the hazards of compounding seemed to be absorbed as, after the building of No 10000, no further compounds were constructed in Britain. For many engineers, the comparative failure of such a high-profile engine could have meant an early retirement but Gresley had displayed his considerable talents in so many other ways and, furthermore, by the time No 10000 had been recalled for rebuilding, his A4s had emerged to provide him with all the moral support he could possibly have hoped for. But externally, No 10000 offered just a little taste of things to come.

CHAPTER 3

Heaters, Boosters, Tanks and Mikados

The late 1920s had been a hectic period for Nigel Gresley. The preparatory work prior to the introduction of the W1 4-6-4 had been very time-consuming, but that locomotive had been just one of many experiments which had been initiated during those years. Since 1912, Gresley's chief assistant had been Oliver Bulleid who was known to have as inquiring a mind as his superior. When it came to probing new areas for development, Bulleid had been as eager as Gresley to explore alternative courses, and much of the experimental

work credited to the boss had derived considerable benefit from the untiring enthusiasm of the assistant. Bulleid was, of course, to get his own command in the future and, when the Southern Railway appointed him as their chief mechanical engineer in 1937, his Gresley-influenced upbringing was no doubt considered an integral part of his CV.

One of Gresley's long-running fields of experimentation involved feed-water heaters, and he had first twiddled with the devices in pre-grouping days. The theory behind the heaters was that

Gresley's early experiments with A.C.F.I. feed-water heaters were conducted in the late 1920s and one of his guinea pigs was an ex-North Eastern Atlantic (by then, LNER No 728). As can be seen from this picture of No 728 at York in 1930, the equipment did little for the locomotive's appearance.

Photo: Rail Archive Stephenson

exhaust steam could be used to raise the temperature of the water before it entered the locomotive's boiler. The use of pre-heated water resulted in greater thermal efficiency and, consequently, a saving on locomotive coal but Gresley usually found that the economies achieved were offset by the additional maintenance costs. Several types of heaters had been tried with varying degrees of success but, during the late 1920s, the French-built A.C.F.I. heater was found to be the most consistent. On the Continent, the A.C.F.I. feed-water heaters were used extensively but they were never to achieve the same level of popularity in Britain. Gresley was, arguably, their staunchest supporter this side of the Channel but even he did not favour their wholesale application. The only LNER locomotives equipped with the A.C.F.I. devices in the 1920s were fifty-five B12 4-6-0s, two C7 4-4-2s and a pair of 4-6-2s.

In the early days of the LNER, the dynamic duo of Gresley and Bulleid had experimented with boosters, the development of which had originated in America. A booster was, in effect, a miniature engine which drew its steam from the locomotive boiler and, from a position underneath the footplate, it would drive the rear trailing axle thereby providing additional power when starting with heavy loads or on steep gradients. In America, boosters were considered invaluable, particularly on some of the country's legendary heavyweight freight trains, but Gresley and Bulleid were more interested in applying boosters to passenger locomotives.

An ex-Great Northern 4-4-2, LNER No 4419, had been the first locomotive which Gresley had fitted with a booster and, in trials in 1923, the results had been remarkable. The 4-4-2 had found no difficulty in starting with an eighteen-coach train of some 535 tons, even on gradients of 1 in 100, and the overall performance had been more typical of an ex-works 4-6-2. The drawback with the boosters was that, as they took their steam from the boiler, their use on anything less than a large-boilered free-steaming locomotive would noticeably reduce the power available from the boiler and the exercise would, therefore, be self-defeating.

In view of the impressive difference the booster had made to No 4419, it was rather surprising that Gresley and Bulleid did not persevere with the idea for use on larger-boilered engines. During the

James Holden's Great Eastern 4-6-0s were classified as B12s after the grouping and, in 1927, three of the class were selected for Gresley's experimental fitting of A.C.F.I. feed-water heaters. Eventually, fifty-five B12s were equipped with the devices and, here, LNER No 8529 is seen modelling the equipment at Stratford on 1 July 1939.

Photo: Rail Archive Stephenson

Although the Lentz rotary valve gear worked well on the D49 4-4-0s, Gresley always kept his eyes open for alternatives and, in 1929, he fitted two ex-GCR four-cylinder B3s with Caprotti valve gear. The results of the experiment showed a saving in coal consumption but not enough to warrant wholesale rebuilding. Two other B3s were fitted with Caprotti gear in 1938 but the outbreak of war the following year put an end to the experiments. LNER No 6166 *Earl Haig* was one of the two ex-GCR engines to receive the Caprotti equipment in 1929 and this picture shows the locomotive at Neasden in September 1932.

Photo: Rail Archive Stephenson

1920s, the only other LNER locomotives fitted with boosters were the massive P1 2-8-2s but, when the length of their trains had to be reduced because of operational problems, the boosters were rendered almost superfluous. However, the P2's boosters were not removed until 1938, three years after No 4419 had lost its own appendage. The patient experiments with boosters occupied a lot of time and resulted in some inventive language, but the first new Gresley design which appeared in the 1930s owed nothing at all to the contraptions.

During the late 1920s, the loadings of suburban trains in the Edinburgh and Glasgow areas had increased to such an extent that the Reid 4-4-2Ts, which the North British Railway had introduced for these duties in 1911, were struggling to cope. Elsewhere on the LNER, Gresley had solved the problem of suburban motive power by the building of twenty-three A5 4-6-2Ts which were virtually identical to a Great Central design of 1911. In Scotland, however, a completely new design was called for and the solution took the form of Gresley's V1 2-6-2Ts, the first of which was

completed at Doncaster in October 1930. When contemplating his design for the V1s, Gresley had been well aware that, down on the Great Western Railway, George Churchward had successfully adapted the design of a mixed traffic 2-6-0 to that of a 2-6-2T and, as Gresley had had considerable success with his K2 and K3 2-6-0s, he used the designs of these as a base for the V1s.

The V1s incorporated the favoured arrangement of three cylinders which, along with the steam chests, were cast in one piece. The cylinders were 16in × 26in and, as all three drove the centre axle, the inside cylinder had to be inclined at an angle of 1 in 8 in order to clear the leading coupled axle. The driving wheels of the V1s were 5ft 8in and the five foot diameter boilers had a working pressure of 180lb. By the autumn of 1931, a total of twenty-eight V1s, Nos 2900–27, had been delivered and all except Nos 2901/11 were despatched directly to Scotland. Of the two rebels, No 2901 was given basic trials on suburban services from Marylebone before being despatched to its intended home of Glasgow, while No 2911 spent seven weeks at Kings

And this is how a new V1 should look; fresh from the paint shop in April 1936, LNER No 455 shows off for the camera. This was one of the first of the class to be fitted with a hopper-style bunker from new. Another variation shown in this picture is the 'hump' casing around the steam-pipes; a redesigned cylinder pattern was introduced in 1938 and so engines built from then onwards had straight pipe covers. When new, No 455 was allocated to Gateshead shed and at the time of Nationalisation it was shedded at Neville Hill. Under BR ownership, it became No 67657 and was rebuilt as a V3 in 1956; withdrawal ensued in December 1962.

Photo: E. R. Morten

Cross shed during the early summer of 1931 during which it turned in some excellent work on services to Hitchin.

The average loading given to No 2911 during its stay in London was around 250 tons but, with the tight timing of services and the climbs beyond Kings Cross, the working was far from easy. Nevertheless, No 2911 was often recorded at speeds in excess of 70 mph once it had reached the more favourable gradients beyond Potters Bar. The crews at Kings Cross were somewhat miffed when the V1 was taken away to join its pals in Scotland, as they had been impressed by the locomotive's capabilities and it was felt that the 2-6-2T had distinct advantages over their usual steeds, the N2 0-6-2Ts, on fast suburban work.

In Scotland, the LNER operated a number of fast suburban services from Waverley station in Edin-

burgh and, on these duties, the V1s soon became as popular with local crews as No 2911 had been at Kings Cross. The Edinburgh-based V1s were divided between St. Margarets and Haymarket sheds and were used mainly on services in Musselburgh, Corstophine and Gorebridge. The sub-sheds at Dunbar and North Berwick usually retained at least one of the V1s apiece.

In the Glasgow area, the sheds at Eastfield, Kipps and Parkhead all received V1s, while the sub-sheds at Arrochar, Balloch, Burnbank, Helensburgh and Lennoxtown normally accommodated one or more of their parent's engines. The LNER's suburban services around Glasgow were fewer than those in Edinburgh, as the LMS offered stiff competition over the former lines of the Caledonian and the Glasgow & South Western Railways in and around the city. But, despite the competition elsewhere in

59

TABLE 3.1: ORIGINAL DIMENSIONS OF THE V1 AND V3 CLASS 2-6-2Ts

	V1 CLASS	V3 CLASS
BUILT:	Doncaster 1930–39	Doncaster 1939–40 (a)
WEIGHTS FULL:	84 tons 0 cwt	86 tons 16 cwt
CAPACITY (bunker):	4 tons (b)	4 tons 10 cwt
(tanks):	2,000 gallons	2,000 gallons
WHEELBASE:	32ft 3in	32ft 3in
WHEEL DIAMETERS (leading):	3ft 2in	3ft 2in
(coupled):	5ft 8in	5ft 8in
(trailing):	3ft 8in	3ft 8in
CYLINDERS:	(3) 16in × 26in	(3) 16in × 26in
BOILER PRESSURE:	180 lb	200 lb
HEATING SURFACES (tubes):	830 sq ft	830 sq ft
(flues):	368 sq ft	368 sq ft
(firebox):	127 sq ft	127 sq ft
(superheater):	284 sq ft	284 sq ft
GRATE AREA:	22.08 sq ft	22.08 sq ft
TRACTIVE EFFORT @ 85% boiler pressure:	22,464 lb	24,960 lb
LNER ROUTE AVAILABILITY:	6	7
BR POWER CLASSIFICATION:	4MT (c)	4MT

Notes:

(a) Most of the V1s were eventually rebuilt as V3s.

(b) Later locos had capacity of 4 tons 10 cwt.

(c) From May 1953, the V1s regraded as 3MT.

Haymarket-based V1 No 2915 is seen at Edinburgh in March 1933. This locomotive eventually became BR No 67615 and, like most of its chums, was rebuilt as a V3.

Photo: E. R. Morten

the area, the LNER had a monopoly of the service to Helensburgh which was one of the most lucrative local commuter runs. During the 1920s, Helensburgh had become one of the more popular dormitory towns for Glasgow and the LNER was adequately aware of the need to woo business passengers travelling to and from the town. Advance revenue from season tickets had to be respected. Consequently, when the first V1s were delivered to Glasgow, the Helensburgh line was considered a priority for their attentions.

Between December 1934 and April 1935, five more V1s, Nos 2928–33, were constructed at Doncaster and all followed their predecessors to Scotland. Doncaster completed a further twenty-eight V1s by October 1936 and, while three were sent to Scotland, the others were despatched to the Blaydon, Heaton and Gateshead sheds on Tyneside.

At first, their duties took them little further than Durham, Seaham, Alnwick and Haltwhistle but they soon expanded their repertoires to include trips to Carlisle and Middlesborough. The railway authorities at Middlesborough were impressed by the V1s and, in 1939, they managed to purloin a pair from Blaydon shed for their own depot. The first twelve V1s which went to the North-East had, like the earlier locomotives, conventional rails on the top of their four-ton bunkers but, from November 1935, a new high-sided hopper-style bunker was introduced and this became standard for the later locomotives. Earlier, in 1934, another differential had appeared and this was the use of the vacuum brake for both locomotive and train, in contrast to the original use of the steam brake for the locomotive and the vacuum solely for the train.

TABLE 3.2: THE V1 CLASS 2-6-2Ts

N.B.: A number of the Scottish-based V1s were usually allocated to sub-sheds such as Balloch, Helensburgh and Hawick but it was not until those sheds received separate codings in 1949 that a differential was officially recorded.

FIRST NO.	1946 NO.	BR NO.	BUILT	REBLT AS V3	SHED ALLOCATIONS			WDN.
					NEW	1/1/48	1/4/58	
2900	7600	67600	9/30	3/56	Eastfield	Eastfield	Eastfield	12/62
2901	7601	67601	10/30		Eastfield	Eastfield	Balloch	1/62
2902	7602	67602	10/30		Eastfield	Eastfield	Eastfield	5/62
2903	7603	67603	10/30		Eastfield	Parkhead	Eastfield	4/62
2904	7604	67604	11/30	11/52	Eastfield	Parkhead	Helensburgh	12/62
2905	7605	67605	11/30	10/53	St. Margarets	St. Margarets	St. Margarets	12/62
2906	7606	67606	12/30	12/52	St. Margarets	St. Margarets	Hawick	12/62
2907	7607	67607	12/30	4/58	St. Margarets	St. Margarets	St. Margarets	12/62
2908	7608	67608	12/30	2/61	St. Margarets	St. Margarets	St. Margarets	12/62
2909	7609	67609	2/31	10/53	St. Margarets	St. Margarets	St. Margarets	2/62
2910	7610	67610	4/31		Haymarket	Haymarket	Haymarket	6/61
2911	7611	67611	4/31	7/53	Parkhead	Parkhead	Parkhead	12/62
2912	7612	67612	4/31	9/53	Parkhead	Parkhead	Parkhead	1/61
2913	7613	67613	5/31	4/56	Kipps	Parkhead	Helensburgh	1/62
2914	7614	67614	6/31	9/58	Eastfield	Parkhead	Helensburgh	7/62
2915	7615	67615	6/31	12/53	Haymarket	Haymarket	Haymarket	12/62
2916	7616	67616	7/31	5/60	Haymarket	Parkhead	Helensburgh	12/62
2917	7617	67617	8/31	10/57	St. Margarets	St. Margarets	St. Margarets	8/62
2918	7618	67618	8/31	9/58	St. Margarets	Stirling	Kipps	12/62

TABLE 3.2: CONTINUED

FIRST NO.	1946 NO.	BR NO.	BUILT	REBLT AS V3	SHED ALLOCATIONS			WDN.
					NEW	1/1/48	1/4/58	
2919	7619	67619	8/31	2/57	Eastfield	Parkhead	Helensburgh	12/62
2920	7620	67620	10/31	7/53	Haymarket	Haymarket	Haymarket	11/64
2921	7621	67621	10/31	12/59	Kipps	Parkhead	Parkhead	12/62
2922	7622	67622	10/31		Eastfield	Parkhead	Helensburgh	3/62
2923	7623	67623	11/31	5/59	Kipps	Parkhead	Parkhead	1/62
2924	7624	67624	11/31	11/52	St. Margarets	St. Margarets	St. Margarets	9/60
2925	7625	67625	12/31	8/53	Eastfield	Parkhead	Helensburgh	12/62
2926	7626	67626	12/31	5/53	Kipps	Parkhead	Parkhead	12/62
2927	7627	67627	12/31	9/53	Parkhead	Kipps	Kipps	8/61
2928	7628	67628	12/34	1/57	Eastfield	Parkhead	Helensburgh	11/64
2929	7629	67629	2/35		St. Margarets	St. Margarets	St. Margarets	5/62
2930	7630	67630	3/35		St. Margarets	St. Margarets	St. Margarets	12/62
2931	7631	67631	3/35		Parkhead	Parkhead	Helensburgh	3/62
2932	7632	67632	3/35	5/57	Eastfield	Parkhead	Helensburgh	12/62
2933	7633	67633	4/35	2/60	Parkhead	Parkhead	Parkhead	12/62
417	7634	67634	4/35	10/40	Heaton	Gateshead	Blaydon	4/62
446	7635	67635	4/35	10/60	Blaydon	Heaton	Heaton	9/63
477	7636	67636	5/35	11/52	Blaydon	Blaydon	Blaydon	11/64
479	7637	67637	5/35		Blaydon	Heaton	Heaton	5/62
481	7638	67638	5/35	1/55	Blaydon	Middlesboro'	Hull B.G.	11/64
484	7639	67639	6/35		Blaydon	Middlesboro'	Gateshead	10/62
486	7640	67640	6/35	12/60	Blaydon	Heaton	Heaton	11/64
487	7641	67641	6/35		Heaton	Heaton	Tweedmouth	10/62
497	7642	67642	7/35	4/60	Heaton	Heaton	Tweedmouth	6/64
498	7643	67643	7/35	5/56	Heaton	Parkhead	Parkhead	11/64
402	7644	67644	8/35	7/58	Heaton	Dunfermline	Eastfield	5/62
414	7645	67645	8/35	11/59	Blaydon	Leeds N.H.	Gateshead	9/63
415	7646	67646	10/35	5/56	Gateshead	Leeds N.H.	Heaton	11/64
416	7647	67647	11/35	12/59	Gateshead	Leeds N.H.	Tweedmouth	1/63
418	7648	67648	12/35	11/59	Heaton	Parkhead	Parkhead	1/62
2897	7649	67649	12/35		St. Margarets	St. Margarets	St. Margarets	7/62
419	7650	67650	1/36	9/58	Blaydon	Stirling South	Stirling South	8/61
422	7651	67651	1/36	5/56	Heaton	Heaton	Heaton	5/64
423	7652	67652	1/36	10/52	Blaydon	Heaton	Heaton	12/63
428	7653	67653	2/36	9/54	Gateshead	Heaton	Blaydon	9/63
440	7654	67654	3/36	10/54	Heaton	Heaton	Heaton	9/63
2898	7655	67655	3/36		Parkhead	Parkhead	Parkhead	3/62
454	7656	67656	3/36	7/52	Heaton	Leeds N.H.	Heaton	12/63
455	7657	67657	4/36	2/56	Gateshead	Leeds N.H.	Gateshead	12/62
461	7658	67658	4/36	1/60	Heaton	Blaydon	Heaton	9/63
465	7659	67659	5/36		Blaydon	St. Margarets	St. Margarets	2/62

TABLE 3.2: CONTINUED

FIRST NO.	1946 NO.	BR NO.	BUILT	REBUILT AS V3	SHED ALLOCATIONS			WDN.
					NEW	1/1/48	1/4/58	
466	7660	67660	7/36	5/56	Gateshead	Kipps	Kipps	2/62
2899	7661	67661	10/36	2/59	St. Margarets	Dunfermline	Parkhead	2/62
404	7662	67662	7/38	2/55	Parkhead	Parkhead	Parkhead	1/63
407	7663	67663	7/38	2/56	Stratford	Norwich	Hull B.G.	9/63
420	7664	67664	7/38		Norwich	Norwich	Eastfield	12/62
424	7665	67665	8/38		Cambridge	Norwich	Kipps	6/61
425	7666	67666	8/38	1/61	St. Margarets	St. Margarets	St. Margarets	2/62
447	7667	67667	9/38	8/58	Stratford	B. Stortford	Eastfield	8/62
448	7668	67668	10/38	5/54	Stratford	Stratford	St. Margarets	12/62
451	7669	67669	10/38	1/43	Stratford	Stratford	Dunfermline	9/61
467	7670	67670	10/38	6/56	St. Margarets	St. Margarets	St. Margarets	8/61
469	7671	67671	10/38		Stratford	Stratford	Eastfield	7/60
472	7672	67672	11/38	3/43	Stratford	Stratford	Dunfermline	12/62
478	7673	67673	11/38		Stratford	Stratford	Gateshead	10/62
480	7674	67674	11/38	8/57	Eastfield	Parkhead	Kipps	12/62
483	7675	67675	12/38	10/42	Stratford	Stratford	Stirling South	12/62
485	7676	67676	12/38	9/60	Stratford	Stratford	Parkhead	7/62
488	7677	67677	1/39	9/58	Stratford	Stratford	Hull B.G.	10/62
489	7678	67678	2/39	10/58	Parkhead	Parkhead	Parkhead	11/64
490	7679	67679	2/39	10/53	Stratford	Stratford	Parkhead	1/62
491	7680	67680	2/39		Stratford	Stratford	Eastfield	12/62
496	7681	67681	2/39	11/57	Stratford	Stratford	Parkhead	11/61

TABLE 3.3: THE V3 CLASS 2-6-2Ts

FIRST NO.	1946 NO.	BR NO.	BUILT	SHED ALLOCATIONS			WDN.
				NEW	1/1/48	1/4/58	
390	7682	67682	9/39	Gateshead	Gateshead	Blaydon	9/63
391	7683	67683	10/39	Gateshead	Gateshead	Heaton	9/63
392	7684	67684	10/39	Middlesboro'	Middlesboro'	Hull B.G.	11/64
393	7685	67685	12/39	Middlesboro'	Middlesboro'	Heaton	12/62
395	7686	67686	12/39	Middlesboro'	Middlesboro'	Hull B.G.	9/63
396	7687	67687	12/39	Gateshead	Gateshead	Gateshead	12/62
397	7688	67688	2/40	Gateshead	Gateshead	Gateshead	12/62
398	7689	67689	2/40	Gateshead	Gateshead	Gateshead	12/62
399	7690	67690	3/40	Gateshead	Gateshead	Gateshead	11/64
401	7691	67691	4/40	Middlesboro'	Middlesboro'	Heaton	11/64

Another twenty V1s were built in 1938/39 and, in common with the 1935/36 batch, they were given whatever odd numbers were currently vacant in the LNER lists. Fifteen of the newest arrivals were allocated to the former Great Eastern section for working, in the main, semi-fast trains from Liverpool Street to Cambridge, Hertford or Bishops Stortford and twelve of the Stratford contingent were fitted with Westinghouse air brakes. A final ten 2-6-2Ts were constructed in 1939/40 but these were fitted with 200lb boilers which raised their tractive effort from 22,465lb to 24,960lb. Because of the differentials, they were classified V3 and they were readily distinguishable from the V1s by having straight steam-pipes instead of elbow pipes. All ten of the V3s were allocated from new to the North-East where they were divided between the sheds at Gateshead and Middlesbrough and the local crews soon came to prefer them to the older A5 and A8 4-6-2Ts despite the fact that the V3s had, on paper, a slightly lower tractive effort.

While the V3s were under construction, it had been anticipated that they would become the standard fast mixed traffic design for the LNER. They were popular and reliable locomotives and the only significant flaw in their design was that, with tanks of only 2,000 gallons capacity, their range was limited. The operating department would have loved to snaffle the V3s for the Southend line but, on the odd occasions that a V1 or V3 had worked on the line, it had been found their water capacity simply wasn't adequate. World War Two brought an understandable hiatus to the development of further V3s and, when peace was restored, Nigel Gresley had been succeeded by Edward Thompson whose own ideas for fast mixed traffic locomotives did not include the V3s.

Under the 1946 renumbering, the V1s and V3s were given LNER Nos 7600–91 in order of seniority, the ten original V3s being, therefore, Nos 7682–91. Before 1946, however, four of the older V1s had been fitted with the larger 200lb boilers and reclassified as V3s. The reboilered locomotives were No 7634 of Gateshead depot and the Stratford trio of Nos 7669/72/75, but they were not to be renumbered again despite their new status. The classes passed intact to British Railways in 1948 and the locomotives received BR Nos 67600–91. Apart

from Middlesbrough-based V3 No 67684 which had carried LNER apple green livery since 1946, the other locomotives merely exchanged their livery of LNER black for one of British Railways black and complete uniformity was achieved when the lone wolf, No 67684 received a black livery in 1949.

The policy of converting V1s to V3s by fitting 200lb boilers was reintroduced in July 1952 and, including the 1940s rebuilds, a total of sixty-three V1s was eventually upgraded. The rebuilding did not, however, result in any renumbering and the last conversion was not undertaken until February 1961, which left just nineteen V1s in an unadulterated condition. In the early 1950s, the Stratford-based V1s and V3s were displaced by Thompson's L1 class 2-6-4Ts and were sent to join their colleagues in Scotland and the North-East. The L1s had first appeared in 1945 and, eventually, totalled one hundred locomotives, but only a few were allocated to the North-East and so the V1s and V3s remained relatively immune in their hideouts until the mass influx of DMUs in the early 1960s. When wholesale dieselisation reared its ugly head it was not, of course, just the North-East area's V1s and V3s which were affected but also the Scottish contingent and, in both areas, their traditional duties rapidly disappeared.

In their last years, the V1s and V3s were relegated to empty stock and parcels duties and some of the Scottish contingent saw out their days on the former Glasgow & South Western lines. Their counterparts in North-East England had slightly longer lives although, by the end of 1964, the only suburban passenger service in the area which was still regularly steam-hauled was between Saltburn and Darlington and, defiant to the last, No 67690 managed to claim its share of turns on the 7.45 am ex-Saltburn until the end of the year. Withdrawal of the remaining V1s had started in July 1960 and the first V3 had gone two months later although, perversely, five V1s were to be converted to V3s after withdrawal of the class had commenced. The last V1, No 67680, had disappeared in December 1962 but the final V3s, Nos 67620/28/46/90/91, all soldiered on until late November 1964.

Work on the design of the V1s had given Gresley a break from his experiments with boosters but in 1931, the year after the first V1s appeared,

The preserved A3 No 4472 *Flying Scotsman* was repainted at Darlington Works in March 1965 and this resulted in the traditional Darlington 'green cylinder' touch. In September 1966, the locomotive was fitted with a second corridor tender in order to circumvent water shortages and, while this was being done, the engine's number was transferred from the cabside to the second tender; the LNER coat of arms replaced the number on the cab. This photograph was taken at Thurlaston on 10 September 1966 and, when compared to the 1964 picture, the livery changes are clearly seen. However, the auxiliary tender which carried the locomotive's number seems to have been given a day off.

Photo: M. John Stretton

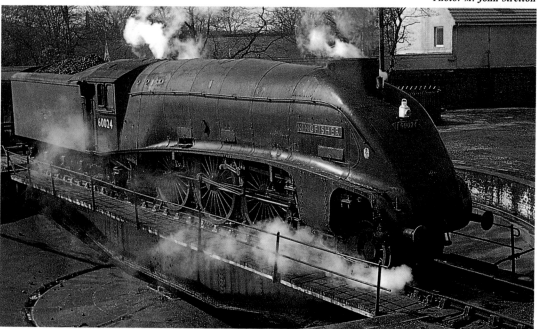

After the end of 1963, the surviving A4s tended to congregate in Scotland. It was 1966 when the last A4s were withdrawn and, in April of that year, No 60024 *Kingfisher* was photographed on the turntable at Ferryhill shed in Aberdeen.

Photo: Jim Winkley

Apart from the overhead electric cables and the locomotive's cabside motif, there are precious few clues as to the date of this picture of No 60009 *Union of South Africa* at Carlisle Citadel. It was, in fact, taken in August 1984, over eighteen years after the engine had been retired by British Railways. The cabside embellishment was applied in 1967 to commemorate the engine's return to steam after its purchase by private owners. The better-known motif is the stainless steel springbok on the casing which was donated by the owner of a South African newspaper and applied in 1954.

Photo: Peter Herring

These two locomotives were built as V3s, as opposed to being rebuilt from V1s. The picture of Blaydon-allocated No 67687 was taken on 7 July 1956 just as it was about to make its way home after a general overhaul at Darlington. By contrast, the other picture shows blatantly-neglected No 67691, which was the last of the class to be built, at Heaton in June 1964. Withdrawal was to follow in five months time and, perhaps, this explains the lack of polish.

Photos: (67687) E. H. Sawford, (67691) Rail Archive Stephenson

Tables 96, 97

Edinburgh and Glasgow to Dumbarton, Craigendoran, Helensburgh and Balloch

Tables 96, 97

Edinburgh and Glasgow to Dumbarton, Craigendoran, Helensburgh and Balloch—Continued.

Table 96

Table 97 — Glasgow to Milngavie

For Steamer Sailings to and from the Coast, via Craigendoran, see Table 69

The services over two of the ex-LNER Glasgow suburban lines are shown in this Scottish region timetable for 1948. These duties were dominated by Gresley's V1 2-6-2Ts.

TABLE 3.4: YEARLY TOTALS OF V1 AND V3 CLASS LOCOMOTIVES

Totals taken at 31 December each year.

YEAR	V1	V3	YEAR	V1	V3	YEAR	V1	V3	YEAR	V1	V3	YEAR	V1	V3	YEAR	V1	V3
1930	9	0	1936	62	0	1942	80	12	1948	78	14	1954	59	33	1960	20	70
1931	28	0	1937	62	0	1943	78	14	1949	78	14	1955	57	35	1961	16	66
1932	28	0	1938	77	0	1944	78	14	1950	78	14	1956	48	44	1962	0	26
1933	28	0	1939	82	6	1945	78	14	1951	78	14	1957	42	50	1963	0	13
1934	29	0	1940	81	11	1946	78	14	1952	72	20	1958	34	58	1964	0	0
1935	50	0	1941	81	11	1947	78	14	1953	62	30	1959	28	64			

thoughts once again turned to the devices. The locomotives which were selected for this next phase of booster experiments were two of Vincent Raven's 4-4-2s, C7 class Nos 727 and 2171, and the booster which was fitted to each locomotive was carried on a bogie. Ingeniously, the locomotive's trailing axle formed the front axle of the bogie while the tender's front axle formed the rear axle of the bogie, and this partly-articulated configuration kept the arguments about wheel notation going for years.

In order to offset the transfer of steam to the pair of 10½in × 14in cylinders of the booster, each locomotive had its original boiler of 170lb pressure replaced by one of 200lb and the grate area increased from twenty-seven to thirty square feet. The booster, the amended wheel arrangement and the larger boilers were more than the classification department required, not just for a sub-class, but a completely new designation. Gleefully, the classification of C9 was created just for the two modified locomotives.

In controlled tests, the boosters enabled the C9s to start trains of almost 750 tons on the level but, in their intended area of operations, level ground was not a common sight. The duties for which the two locomotives had been earmarked were between Edinburgh and York where the weights of the trains and the tough gradients presented a challenge for any locomotive. The stiffest test on that route was on the Cockburnspath bank in north Berwickshire and, although the C9s managed to haul 400-ton trains to the summit, it was not an easy task for them. The design of the boosters fitted to the C9s had been stretched to the limit so that they would

cut in when the speeds of the locomotives dropped to 30 mph. This had involved considerable ingenuity as the original American boosters could not cut in until speeds dropped to 15 mph.

The text book did not, however, account for the gradients in southern Scotland. By the time a locomotive had been slowed down to 30 mph on its ascent of Cockburnspath, a burst of additional power was, by then, a little too late. The problem of designing a booster which would cut in at higher speeds was never to be completely resolved, and so the use of the devices on the LNER's express locomotives was not pursued. Despite the disappointment with the C9s, Gresley was well aware that, when it came to starting a heavy train on the level or running at very low speeds, the benefits of boosters had been seen time and time again. Therefore, when the next LNER locomotive received the booster treatment, it was a heavy shunting engine.

In January 1932, a booster was fitted to S1 class 0-8-4T No 6171 which was one of four locomotives that had been built in 1907/08 by Beyer Peacock to a Robinson design for the Great Central Railway. The three-cylinder 0-8-4Ts had been conceived specifically for heavy shunting in the extensive hump marshalling yard which had been opened in 1908 at Wath in the West Riding of Yorkshire and, on the 1 in 146 gradient leading to the hump, the S1s were allowed up to 720 tons unaided. Here were locomotives which seemed perfect for being fitted with boosters as their duties required not speed but strength.

The fitting of a booster to No 6171 was not entirely straightforward as the locomotive had, by the very nature of its duties, to be able to operate

After mixed success with his experiments with boosters, Gresley fitted S1 class 0-8-4T No 6171 with one of the devices in 1932. It worked a treat. Consequently, he ordered two new S1s and the second of these, LNER No 2799, is seen at Mexborough shed in July 1933. After a lifetime spent in the heady delights of hump marshalling yards, the locomotive was retired as BR No 69905 in 1957. The coupling rods joining the trailing wheels can be clearly seen but these were removed when the booster was dispensed with in 1943. The cut-away tops to the tanks had first been applied to Gresley's J50 0-6-0Ts in pre-grouping days and the idea behind the 'missing' segment was that the crews' visibility would be improved when undertaking shunting duties. With locomotives as large as the S1s, it was a sensible bit of design work.

Photo: Rail Archive Stephenson

either chimney- or bunker-first. Gresley's previous boosters had been unidirectional, as the locomotives to which they had been fitted would have been expected to carry out their tasks facing forwards but, in the case of No 6171, a reversible booster had to be applied. At the same time as the locomotive received its booster, it was fitted with a superheated boiler and, as a result of some clever thinking, the wheels of its rear bogie were coupled so that the booster could provide the maximum additional adhesion possible. In its new form, No 6171 showed off by shunting 1,000 ton loads over the hump even with the hindrance of West Riding rain on the rails. Without the added paraphernalia, the S1s had had a commendable tractive effort of 34,523lb but the rebuilt locomotive could, with the aid of its booster, muster as much as 46,896lb.

The rebuilding of No 6171 was considered to be an unqualified success and, instead of treating the

three other S1s in a similar fashion, Gresley went one better and ordered the construction of two new 0-8-4Ts to the same specifications as the rebuild. The new arrivals were built at Gorton in 1932 and this must have been a novelty for the staff at the works as, between 1928 and 1938, they were the only locomotives to be built there. The requirement for the new engines was not for shunting at Wath but for the marshalling yard which had recently opened at Whitemoor, near March in Cambridgeshire.

Carrying Nos 2798/99, the two new engines were welcomed by the classification department. Mechanically, Nos 2798/99 were almost identical to the rebuild, No 6171, but cosmetically, the new pair followed the lead set by the J50 0-6-0Ts way back in 1914 by having the front ends of their tanks cut away in order to improve the visibility from the cab. This opened the door to divide the six loco-motives of the class into three sub-classes. The

TABLE 3.5: ORIGINAL DIMENSIONS OF THE S1/2 CLASS 0-8-4Ts

BUILT:	Gorton 1932 (a)
WEIGHT FULL (locomotive):	104 tons 5 cwt
CAPACITY (bunker):	6 tons
(tanks):	2,680 gallons
WHEELBASE:	31ft 2in
WHEEL DIAMETERS (coupled):	4ft 8in
(bogie):	3ft 2in
CYLINDERS (locomotive):	(3) 18in × 26in
(booster):	(2) 10in × 12in
BOILER PRESSURE:	180 lb
HEATING SURFACES (tubes):	885 sq ft
(flues):	464 sq ft
(firebox):	151 sq ft
(superheater):	242 sq ft
GRATE AREA:	26.24 sq ft
TRACTIVE EFFORT @ 85% boiler pressure:	34,523 lb
(booster @ 72.5%):	12,373 lb
LNER ROUTE AVAILABILITY:	8
BR POWER CLASSIFICATION:	7F (6F from May 1953)

Note:

(a) Original members of the S1 class built by Beyer Peacock 1907 – 08.

three unaltered ex-Great Central machines became S1/1, the solitary rebuild became S1/2 and the two new engines became S1/3. In the early 1940s, two of the unrebuilt S1s of 1907/08 were superheated and the third followed suit in 1951. Under Edward Thompson, the boosters were removed from Nos 2798/99 and 6171 in the 1940s and, in the 1946 renumbering, the six 0-8-4Ts became Nos 9900 – 05.

At the time of Nationalisation, Nos 9902/03 were serving at March and the other four were still allocated to Mexborough shed for their duties at Wath. The pair at March were displaced by diesel shunters in 1949 and were transferred, initially, to Frodingham but they found their way back to Mexborough the following year. The S1s were specialised locomotives with a very limited sphere of operations and, when diesel shunters appeared at Wath in 1953, there was little alternative work available for them. Two of the class had unsuccessful spells at Immingham, two later went to Frodingham but the others were sent to Doncaster shed and were rarely steamed again. The Frodingham

pair, BR Nos 69901/05 (LNER No 6171/2799, later Nos 9901/05), were the last survivors and retired simultaneously in January 1957 after being replaced, not by diesel shunters, but by J50 0-6-0Ts.

The idea for tampering with the S1 0-8-4Ts was first discussed in 1931 and, that same year, Gresley got to work on another idea for conversion of tank locomotives. Nobody could accuse Gresley of doing things by halves, as the conversion scheme which he drew up in 1931 did not affect just one or two locomotives but an entire class and, this time, boosters did not feature anywhere in his plans. In the Newcastle and Middlesbrough areas, fast suburban and secondary passenger services had long been monopolised by two classes of former North Eastern Railway locomotives, Worsdell's 0-4-4Ts and Raven's 4-4-4Ts. The two-cylinder 0-4-4Ts had originated in 1894 and a total of 110 had been built by 1901 but, understandably, the more arduous duties around the North-East were, by the 1930s, entrusted to the more modern three-cylinder 4-4-4Ts, forty-four of which had been built between 1913 and 1922. In 1931, however, Gresley decided to rebuild the whole lot of 4-4-4Ts as 4-6-2Ts.

The decision to convert the 4-4-4Ts was partly a belated, but calculated, public relations exercise. In 1927, an accident involving one of the Southern Railway's River class 2-6-4T at Sevenoaks reopened the long-standing national debate about the wisdom of using rear bogie designs for fast running. As a result of the publicity generated by the Sevenoaks accident, public confidence in rear bogie engines was not all it could be and, considering that the 4-4-4Ts were prone to wheelslip and a degree of instability at speed, Gresley felt that the conversion of the locomotives to 4-6-2Ts would not only improve their usefulness but also their public image.

Gresley had already seen the benefits of the additional adhesion offered by 4-6-2Ts in the North-East as, between 1923 and 1926, a batch of twenty-three A5 4-6-2Ts had been constructed to a Robinson design of 1911 for use in the area. In July 1931, H1 4-4-4T No 2162 was despatched to Darlington for surgery and, between January 1933 and August 1936, the other forty-three members of the class followed suit. Their new classification was A8. Despite being subjected to what seemed like a complete transformation, the innards of the 4-4-4Ts were hardly altered during the rebuilding and, apart from the new wheel arrangement and an increase in weight of just over two tons, their vital statistics remained almost unchanged. Contemporary observers commented that, with their outside cylinders and inside valve gear, the A8s were still, quite unmistakably, Robinson locomotives. The class passed intact into the hands of British Railways in 1948 to carry Nos 69850–94 and, apart from six which were allocated to Botanic Gardens depot in Hull, all were still at work in North Yorkshire and the North-East. The last eleven survivors were withdrawn in June 1960.

The rebuilt S1s and the A8s materialised as a result of Gresley's determination to improve his stock of locomotives but neither class could justifiably claim a place in his portfolio of designs. After the introduction of the V1 2-6-2Ts in 1930, the next few years saw acute financial depression throughout the country and this recession affected the LNER no less than Britain's other railways. Expenditure on the development of new locomotive designs had to be kept to a minimum and, although plans were afoot on the LNER for a new type of express passenger locomotive, Gresley had to be

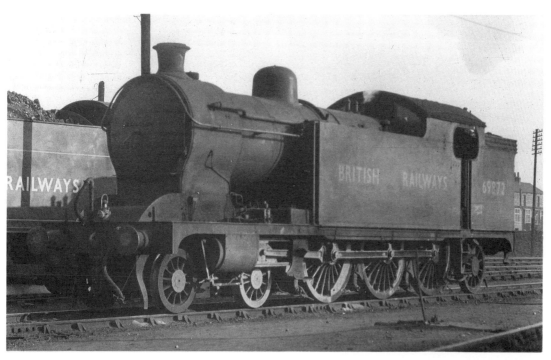

Once upon a time, the H1 4-4-4Ts were popular engines on secondary passenger services, particularly in their native North-east. However, Gresley rebuilt the whole lot as 4-6-2Ts between 1931 and 1936 and, consequently, they were reclassified as A8s. The picture on the previous page shows LNER No 1521 (later BR No 69890) at Darlington in 1937, the top photo is of No 69872 in early BR livery at Darlington in March 1950 while the bottom photo shows a well-tended No 69892 approaching Darlington with a local train in 1955. The only significant difference between the engines, apart from their liveries, is that LNER No 1521 is fitted with a cage bunker. Nos 69872/90/92 were all retired in 1958, their final months being October, January and November respectively.

All photos: Rail Archive Stephenson

The Claud Hamiltons were, arguably, the most famous engines to be built by the Great Eastern Railway. During the 1930s, Gresley performed some extensive modernisations to these elegant machines and two examples are seen in this pair of pictures. The photograph of No 62569 was taken on 1 July 1954 at St. Ives station on the March to Cambridge branch; like the locomotive, the station has long since departed this world. The original footplating of the locomotive is clearly visible and this can be compared with the modified footplating of No 62576 which was photographed at Cambridge on 9 May 1955. One internal difference between these two Clauds is that No 62569 retained its original slide valves while No 62576 had been rebuilt with 9½ inch piston valves.

Both photos: E. H. Sawford

content with spending much of his time and little of the LNER's money on updating older designs.

One large section of the LNER's motive power stock which had caused Gresley frequent problems was that of the express passenger locomotives which had been inherited from the Great Eastern at the grouping, the majority of which had been designed by James Holden, his son Stephen or Alfred Hill. Although the Great Eastern's engines had matched those of any other company in the early 1920s, they were often hard-pressed to cope with the demands of the 1930s. The Great Eastern's crack express locomotives had been James Holden's famous Claud Hamilton 4-4-0s and his son's 1500 class 4-6-0s which, under LNER ownership, had been classified as D14/D15/D16 and B12 respectively but, apart from Gresley's experiments with feed-water heaters on some of the B12s, little attention had been paid to the others. One of the factors which inhibited work on former Great Eastern express locomotives was that they had been built for the ungenerous weight restrictions imposed by the permanent way, and this problem had already manifested itself by the limitations which had had to be applied to the design of the B17

Sandringhams in 1928. Anything that added weight to an East Anglian locomotive had, until 1930, been strictly taboo.

The enforced lull in new developments in the 1930s gave Gresley the opportunity to see how he could improve the former Great Eastern machines and, by this time, several of the lines in East Anglia had been improved so that heavier axle weights could be tolerated. In 1932, Gresley got to work on B12 No 8579 and fitted it with a larger boiler, a round-topped firebox and modified valve gear, all of which added 6¼ tons to its previous weight of 63 tons. Apart from the improvement in performance, No 8579 showed a conspicuously reduced appetite for coal and so the order went out to start rebuilding other B12s.

In all, fifty-four of the eighty-one B12s received the treatment at Stratford works and, when they returned to business, they behaved like born-again racers. On the routes in East Anglia, loadings were not often in excess of 300 tons and gradients were rarely a problem, but this did not mean that conditions were idyllic. Timings were very tight and, furthermore, only a relatively small proportion of express services in the area were scheduled to run

And here's another Gresley rebuild. This picture shows B12/3 No 61579 leaving Prittlewell with the 1.13pm Liverpool Street to Southend stopper on 2 July 1955. The B12s had originated on the Great Eastern and, by the early 1930s, they were not exactly in their prime. Gresley extensively rebuilt one of the class in 1932 and the result was so startling that fifty-four of the eighty-one B12s were given similar treatment.

Photo: LCGB/Ken Nunn Collection

Gresley's rebuilding of the B12s gave them new leases of life and, even a decade or more after the work was done, the locomotives were still considered suitable for express workings on the former-Great Eastern network. This shot of Stratford-based No 61568 was taken on 5 November 1950 and shows it passing Chadwell Heath at the head of the Day Continental (the 9.25am Liverpool Street to Parkeston).

Photo: LCGB/Ken Nunn Collection

non-stop and so acceleration was as essential as a high top speed. The rebuilt B12s coped admirably. Naturally, the classification department could not let the modifications go unrecorded and the revamped locomotives became B12/3 while the unrebuilt ones became B12/1.

While the B12/3s were left to perform heroics on familiar territory, the B12/1s were gradually transferred to the former lines of the Great North of Scotland Railway where any 4-6-0 with a modest axle weight, no matter how well used, was positively welcomed. At Nationalisation, both variations of the B12s were mixed together as Nos 61500 – 80 without any distinct numerical differentiation. Under the tender loving care of British Railways, the Scottish contingent were all withdrawn within five years but the English brigade staved off extinction until 1960.

Late in 1932, it was the turn of the celebrated Claud Hamilton 4-4-0s to be transformed at Stratford. The treatment meted out to the Clauds was far

more extensive than that received by the B12s. New boilers, fireboxes, cylinders and piston valves constituted, in effect, a renewal rather than a conventional rebuild. At first, ten Clauds were selected for the operation and the first to come round from the anaesthetic was No 8848 which reappeared from the works in January 1933 and it was followed one month later by none other than No 8900, *Claud Hamilton* itself.

Further Clauds were rebuilt between 1936 and 1939, some with new piston valves, others with round-topped fireboxes and a few without their coupling rod covers, while thirteen members of the class escaped any treatment at all. Quite astoundingly, a potential red letter day was missed by the classification boys and just three variations were officially acknowledged: D15 for the unrebuilt engines and D16/2 and D16/3 for the doctored ones. Unsurprisingly, the Gresley rebuilds were to have longer lives than their untreated counterparts and, while the D15s all disappeared soon after

Nationalisation, the last four of the rebuilds survived until 1959.

One of the new round-topped boilers which had been destined for a D16 was waylaid in October 1934. Instead of taking up residency on the frames of the 4-4-0, the boiler found itself sitting on the chassis of one of Alfred Hill's ex-Great Eastern 0-6-0 freight locomotives. Gresley had been delighted at the transformation that had been made to the B12s and the D16s and he saw no reason why similar treatment should not work the same wonders for goods engines. He was not disappointed with the first results and so, between 1934 and 1939, not only were all twenty-five J19 0-6-0s reboilered but also the ten members of the J18 class. On this occasion, the classification department lost out as, instead of having new variations to play with, the reboilering of the J18s rendered them so similar to the J19s that the former classification was erased from the books.

During the early 1930s, the relative tedium of updating older designs was partly offset by design work for a new type of express passenger locomotive. Gresley's 4-6-2s were putting in some highly impressive performances but even those machines tended to struggle on the tortuous slog between Edinburgh and Aberdeen. The southbound run over that route was the most difficult and the official loading limit for the A3s was 420 tons but, with the introduction of third-class sleeping cars, the weight limit was sometimes exceeded and, therefore, double-heading had to be employed. All railway companies tried to avoid the expense of double-heading as far as possible but Gresley was, if anything, even more opposed to the practice than most of his contemporaries.

There was also an element of pride involved, as Gresley and the LNER had established a joint reputation for producing some of the most powerful locomotives in Britain and, understandably, it hurt to admit that the designs could not keep abreast of all the demands of the operating department. When the idea for newer, more powerful locomotives for the Aberdeen route was raised, it created not a little controversy. In practice, by no means all of the southbound trains on the Aberdeen run exceeded the 420 ton limit imposed on the A3s and a number of contemporary commentators were quick to remark

that it seemed extremely extravagant to plan new locomotives for a very restricted requirement. The LNER, however, fully appreciated the benefits of publicity and prestige. When the W1 4-6-4 had first appeared, the company had hardly been out of the headlines and, furthermore, as Gresley and his staff had patiently bided their time during the recession, it was felt that the authorisation for the new machine would be an important morale-booster for the design lads.

The answer to the problem on the Aberdeen run took the form of the giant 2-8-2, No 2001 *Cock O' The North*, which was completed at Doncaster in May 1934. Classified P2, it was the first eight-coupled express passenger locomotive in Britain. Its name had been forcibly purloined from William Reid's North British Railway 4-4-2 No 903 (LNER No 9903) whereas the choice of its number had involved less work from the screwdriver. The Wilson Worsdell locomotive which had previously carried LNER No 2001 had been withdrawn in 1931 and, as it had been Britain's first express passenger 4-6-0, it was considered appropriate that the number should be reused on another innovative machine.

The new locomotive incorporated Gresley's pet three-cylinder arrangement and, with driving wheels of 6ft 2in and a tractive effort of 43,462lb, there were no doubts at all that it would handle trains of 550 tons on the Aberdeen route with ease. No 2001 certainly looked the part, but not just because of its ample proportions. At its front end, it was fitted with integral smoke deflectors which were very similar in design to those used on the W1 4-6-4, and the effect on No 2001 was a sleek, semi-streamlined appearance. The impression of streamlining was enhanced by the wedge-shaped front of the cab, and this feature was incorporated, not just for cosmetic appeal, but also as an aid to smoke-deflection.

No 2001 was fitted with Lentz rotary cam gear as had been tried on a number of the D49 4-4-0s. Gresley included an A.C.F.I. feed-water heater in the design of No 2001 and this was, perhaps, surprising in view of the mixed results he had previously obtained with the devices. The A.C.F.I. heater was not the only French-influenced feature incorporated in the design of No 2001, as its

The front end of P2 2-8-2 No 2001 *Cock O' The North* bore a marked resemblance to the water-tube 4-6-4 No 10000 and, perhaps, that is the reason why the arrival of the P2 was treated with a little scepticism in professional circles. This picture shows the powerful-looking creature at Kings Cross on 11 May 1935 and the A.C.F.I. feed-water equipment can be seen on the running plate.

Photo: LCGB/Ken Nunn Collection

draughting arrangement was similar to the system perfected by André Chapelon on the Paris-Orleans Railway. The locomotive's exhaust was discharged through twin blastpipes and a double chimney and, although this idea had first been devised by the Finnish engineer, Kylala, in 1919, it had taken Chapelon's perseverance to develop it. The double chimney design became universally known as the Kylchap system after Messrs. Kylala and Chapelon.

After completing its preliminary tests, No 2001 performed its first service trials southwards from Doncaster and, on express duties to Kings Cross, the locomotive regularly reached the speed of 85 mph which had previously been unheard of for an eight-coupled design. Other trials were regularly undertaken between Kings Cross and Peterborough, but the most remarkable demonstration run took place on 19 June 1934 when No 2001 was put in charge of a 650-ton train from Kings Cross to Barkston, just to the north of Grantham, and back. The run did not include any severe gradients but, on the

northbound trip, the climb to Stoke summit, north of Corby, was a lengthy one which peaked at 1 in 178 but it was only on the final approach to the summit that the heavily loaded locomotive dipped below 60 mph. The following month, No 2001 was despatched to its intended stomping ground of Scotland in time to get the feel of the road before the demands of the peak-season holiday traffic.

Doncaster works presented No 2001 with a brother in October 1934 when No 2002 *Earl Marischal* was delivered. The junior locomotive differed from its senior in having conventional piston valves and Walschaerts gear instead of the Lentz rotary equipment, but this change did not reflect any feeling of dissatisfaction with the Lentz gear. Gresley simply felt that it would be useful to draw comparison between the two types. A further differential between the pair of P2s was that No 2002 was fitted with a Davies & Metcalf exhaust steam injector instead of the A.C.F.I. feed-water heater but, although the injector itself was most

TABLE 3.6: ORIGINAL DIMENSIONS OF THE P2 CLASS 2-8-2s

BUILT:	Doncaster 1934 (four more built 1936)
WEIGHTS FULL (locomotive):	110 tons 5 cwt (a)
(tender):	55 tons 6 cwt (a)
TENDER CAPACITY (water):	5,000 gallons
(coal):	8 tons
WHEELBASE (locomotive):	37ft 11in
(tender):	16ft 0in
WHEEL DIAMETERS (leading):	3ft 2in
(coupled):	6ft 2in
(trailing):	3ft 8in
CYLINDERS:	(3) 21in × 26in
BOILER PRESSURE:	220 lb
HEATING SURFACES (tubes & flues):	2477 sq ft
(firebox):	237 sq ft
(superheater):	776.5 sq ft
SUPERHEATER:	43 elements
GRATE AREA:	50 sq ft
TRACTIVE EFFORT @ 85% boiler pressure:	43,462 lb

Notes:

(a) The later streamlined P2s (see Chapter Five) had weights of: locomotive: 107 tons 3 cwt; tender: 57 tons 18 cwt.

In this picture of P2 2-8-2 No 2002 *Earl Marischal*, the additional Southern-style smoke deflectors can be seen and, considering the exhaust generated by the locomotive for its departure from Kings Cross, perhaps those extra deflectors were a good idea. The photograph was taken on 18 May 1935 and the train is the 4.00pm to Newcastle.

Photo: LCGB/Ken Nunn Collection

satisfactory, its softer exhaust blast meant that smoke was not lifted clear of the driver's field of vision. To augment its original integral deflectors, additional side plates similar to those used by the Southern Railway were fitted to No 2002.

In December 1934, No 2001 tentatively hauled a short coal train to Harwich but, much to the relief of the ex-Great Eastern civil engineers, this was not the prelude to a regular working. The trip to Harwich was to enable the locomotive to be shipped to France in order to undergo static trials on the testing plant at Vitry-sur-Seine, and the wagons of coal were for the engine's own consumption as the LNER did not want No 2001 to be adversely affected by a foreign diet. The reasons behind the visit to France were twofold. The first was to show gratitude to French engineers who had contributed, knowingly or not, to the design of No 2001, and the second reason was that Gresley had often argued the case for a similar testing plant in Britain.

The French jaunt lasted for three weeks, during which No 2001 performed controlled trials on the open road between Tours and Orleans. Unfortunately for the LNER's publicists, the vacuum brake fitted to No 2001 was not compatible with the Westinghouse system used in France and so the locomotive could not show off by hauling passenger trains. Instead, the test loads consisted of nothing more glamorous than three dead French locomotives. During the trials, the French driver and fireman expressed astonishment at the diminutive proportions of the British coal shovel and they suggested that, if a proper man-size shovel were to be used on the locomotive in Britain, its firing could be significantly improved. Gresley thanked the crew for the idea but, mindful of the strength of the trades unions on the LNER, he treated the suggestion with diplomatic amnesia.

Throughout the stay in France, public relations were helped enormously by Oliver Bulleid who surprised one and all with his ability to speak French fluently, and the linguistic exchange became reciprocal when Bulleid taught his hosts some new English words after the axle-boxes of No 2001 started to overheat with disconcerting regularity. On its return to Britain, the first port of call for No 2001 was Doncaster works so that the axle-box problem could be investigated thoroughly. The

trouble turned out to be a fractured bearing and, after the repairs were completed, the locomotive was returned to Scotland to await the arrival of its younger brother.

It was spring 1935 before No 2002 completed its protracted service trials on the seemingly obligatory route between Doncaster and Kings Cross and, when the pair of P2s were eventually in action between Edinburgh and Aberdeen, they left no doubts at all about their ability to haul lengthy passenger trains. This did not, however, provide the great success story that had been envisaged, as the operating department seemed not to have learned any lessons from the P1s of 1925. Few of the stations and sidings in Scotland could accommodate trains of more than fifteen coaches and, when the P2s were in charge of longer trains, it usually meant double stops at stations. This not only made a mockery of their schedules but it also had a knock-on effect for other trains in the same or adjacent sections.

The problem of overheated axle-boxes started to reappear and, although the cause could not be positively identified, the concensus of opinion was that the curving nature of the track northwards from Edinburgh, in particular around Montrose, placed excessive stress on the long wheelbases of the locomotives. Consequently, it was decided to make life easier for the P2s by changing the locomotives at Dundee rather than having them run through between Edinburgh and Aberdeen. While that solution might have been sound in theory, in practice it resulted in the locomotives spending much of their time waiting for turnrounds and, furthermore, it significantly reduced the daily mileages of each locomotive. The waits and the low mileages made for inefficient use of the P2s and, as a result, their fuel consumption was excessively high but, unfortunately, the term 'self-defeating' was not altogether foreign to the LNER's design staff.

A further problem with the P2s was that they were prone to derailment at difficult sets of points in shed yards. The simple solution to this was to allocate them specific 'safe' roads in the yards but, when the movements in an entire shed were dictated by the needs of just one or two locomotives, it did not make for easy management of a depot.

The distinctive Gresley parentage of P2 No 2001 *Cock O' The North* is clearly evident in this excellent picture. Apologies, dear reader, but details of the date and location of this photograph are not known. A seemingly daft but, nevertheless, interesting little exercise is to lay a picture of this locomotive alongside one of a BR Standard Britannia and to flit from one picture to the other with slightly closed eyes; to a swift glance, the similarity is uncanny. Should you have bought this book on April 1st, please rest assured that no joke is intended!

Photo: Rail Archive Stephenson

Even worse, there were rumours that the P2s were responsible for spreading the track on the main line but, as the accusers could not state beyond reasonable doubt that the offending machines were in fact the P2s, the threat of a hasty demise was lifted.

Despite the faults of the P2s and their mismanagement, there was no doubt about their power. The LNER was a publicity conscious company and it made the most of the fact that the P2s were, on the basis of tractive effort, the most powerful passenger locomotives in Britain. Not only that, but they were truly impressive-looking machines which, taking the operating problems aside, were a publicist's dream. Four more P2s were to be built in 1936, but these were to display distinct cosmetic differences which were prompted by the phenomenal success which was just around the corner for Nigel Gresley.

CHAPTER 4

The Masterpieces

Gresley was by no means an egotistical man but, during the early 1930s, his W1 4-6-4 had caused him a degree of embarrassment. This had been followed in the early part of 1935 by the realisation that his P2s were, if anything, too powerful for their own good. Although he had nothing to prove, Gresley was not oblivious to the fact that, while the practice of rebuilding older locomotives to give them a further quarter-century life expectancy might make excellent economic sense, it was not the stuff that maintained reputations. Gresley's frustrations were partially alleviated by the ton-up exploits of A3 No 2750 *Papyrus* on 5 March 1935 and also the preparatory work for a new class of 4-6-2s and, had he been able to foresee the all-round acclaim which his new locomotives were to receive, he would have been rather more content.

The need for new 4-6-2s arose because of the decision to introduce a high-speed streamlined train between London and Newcastle in September 1935. Its inauguration was planned to coincide with the twenty-fifth anniversary of King George V's coronation and was to be called, appropriately, the Silver Jubilee. From the outset, it was decided that not only the trains but also the locomotives themselves would be streamlined and, while this directive might have panicked less-experienced designers, it was, by now, old hat to Gresley. His first venture into streamlining had been with the W1 4-6-4 and, although the contraption's mechanical behaviour had been likened to the away form of Doncaster Rovers, its external appearance was certainly a crowd-pleaser. Similarly, the P2 2-8-2s may have gone on to cause occasional bouts of apoplexy in the operating department but, when

they first appeared, their semi-streamlined looks kept a good few photographic retailers in profits.

Gresley had kept a close eye on developments in Germany where a streamlined two-car diesel unit regularly grabbed the headlines because of its speedy exploits between Berlin and Hamburg, and it was the streamlining aspect which interested him. It was not that Gresley had anything against the use of diesels but he considered that, for long-distance main-line duties from Kings Cross, steam would prove far superior. Apart from Gresley and a few other eminent British engineers, it was difficult to find many people in this country who took the German diesel unit too seriously as, although the Teutonic name of 'Der Fliegender Hamburger' sounded pretty impressive, the English translation of 'The Flying Hamburger' removed much of the gloss.

By the spring of 1935, many of the final details of the new 4-6-2s had been thrashed out and the locomotives were designated A4s. Mechanically, their design had started life as a 4-6-2 version of the P2s but, instead, finished up more like an A3, complete with 6ft 8in driving wheels, but with a 250lb boiler and 18½ in × 26in cylinders. However, externally, things were very different indeed to anything that had gone before. The idea for the streamlining might have been inspired by the German diesel units, but Gresley leant more heavily on the Bugatti design which was used successfully for petrol railcars operating between Paris and Deauville in France. Gresley's adaptation of the Bugatti design resulted in the most comprehensively streamlined design so far seen in Britain.

Not all of the external features of the A4s were

This superb picture was taken at York on 3 July 1988. The magnificent trio of A4s was lined up as part of the 'Mallard' celebrations and, quite fittingly, the locomotive nearest the camera is No 4468 *Mallard* itself. Behind it seems to be No 2509 *Silver Link* but, underneath the impressive livery, is really a repainted No 60019 (LNER No 4464) *Bittern*. At the rear, the A4 which has picked up its skirts is No 4498 *Sir Nigel Gresley*.

Photo: Robert Falconer

Most appropriately, A4 No 60007 *Sir Nigel Gresley* was preserved. Although it was restored to Garter blue livery and was given its old number, 4498, a complete return to its original condition was vetoed for reasons which are explained in the text. In other words, if you're browsing through this book in a shop, you'll just have to buy the tome to find out precisely why. Here, No 4498 is seen entering a conveniently placed patch of light at Harbury on its return to Marylebone in September 1988.

Photo: P. Chancellor

It was hardly surprising that, during their final years, the A4s were in regular demand for enthusiasts' specials. This picture shows No 60024 *Kingfisher* at Weymouth shed on 26 March 1966 after it had worked a special from Waterloo; at the time, the locomotive's home depot was a little distance away at St. Margarets in Edinburgh.

Photo: E. H. Sawford

the result of intensive midnight pow-wows or sparks of miraculous ingenuity. The potential problem of the streamlined front end causing restricted access to the smokebox was solved by the use of a remarkably simple handle-operated door, the inspiration for which came to a junior draughtsman when he observed a Doncaster Corporation dustcart being loaded. It was a similar story with the actual lines of the front-end streamline casing. The casing curved in a graceful sweep of twelve foot radius and was, later, to be greatly admired but, although the A4s and all subsequent streamlined LNER locomotives were to have the same front-end curves, the use of a twelve foot radius was the result of a draughtsman's myopia. The original design had stated, quite distinctly, a curve of fourteen foot radius but by the time the mistake was discovered, it was too late for it to be rectified.

Construction of the locomotives was started on 26 June 1935, and the first to be completed was No 2509 which emerged from the works on 6 September. The locomotive was named *Silver Link*

and its contemporaries, Nos 2510/11/12 were given the names of *Quicksilver*, *Silver King* and *Silver Fox* respectively. The common theme of 'Silver' was not just down to their debut in Jubilee year but also because of their liveries which, like that of the streamlined coaching stock which had been built specially for the new London to Newcastle service, comprised different shades of silver-grey. The colour scheme had previously been applied experimentally to two other locomotives but neither could have been considered as competitors for the forthcoming express services. One was an ex-Great Northern 0-6-0ST which had been built by Patrick Stirling in 1891 and the other was a J50 0-6-0T.

The first test run given to No 2509 took place on 27 September 1935 and it involved taking a seven-coach train of 220 tons from Kings Cross to Grantham and back again. Any doubts about the wisdom of using a locomotive which had left the workshops only three weeks previously were soon dispelled as, on the outward trip, No 2509 maintained a speed of over 100 mph for forty-three

The inaugural run of the Silver Jubilee took place on 30 September 1935 and the return left Kings Cross for Newcastle at 5.30pm. On that momentous first day, Gresley A4 No 2509 *Silver Link* was photographed passing Potters Bar on the northbound run with its unmistakable train of seven purpose-built coaches.

Photo: LCGB/Ken Nunn Collection

Predictably, the quartet of Silver A4s monopolised the Silver Jubilee for a long time and, here, No 2510 *Quicksilver* is seen at the head of the train at Newcastle on 14 May 1936.

Photo: H. C. Casserley Collection

miles and twice peaked at 112.5 mph. The 105½-mile trip to Grantham took just eighty-eight minutes. Three days later, on 30 September, the inaugural Silver Jubilee left Newcastle at 10am and, with just one stop at Darlington, it arrived at Kings Cross at 2pm; the return left Kings Cross at 5.30pm. Until No 2510 *Quicksilver* was fully completed two weeks later, No 2509 took charge of all Silver Jubilee services and, despite covering over 6,000 high-speed miles, no mechanical problems were encountered. No 2911 *Silver King* was delivered in November and was allocated to Gateshead shed to act as a standby, whereas No 2912 *Silver Fox* joined the original pair at Kings Cross shed.

The Silver Jubilee service did not require a change of engine crews en route but, nevertheless, the first A4s all had corridor tenders which were, naturally, streamlined. The fact that the Silver Jubilee did not run at weekends released the A4s for other duties and it was understandable that the LNER was quick to use them on the Flying Scotsman between Kings Cross and Edinburgh; the loadings on that train were always far greater than the

seven-coach Silver Jubilee but the locomotives displayed ample strength as well as speed. The addition of streamlining to the tenders reduced their capacities from nine tons of coal to eight, although this was rather close to the margin for comfort on non-stop runs to Edinburgh.

The only persistent grumbles about the A4s came from the purists among the ranks of amateur enthusiasts, who considered that their external appearances were not how real steam engines were meant to look. As for the rest of the general public, the quartet of A4s were a tremendous hit, particularly as their full streamlining fitted perfectly with the contemporary fad for anything that smacked of art deco. The aura which surrounded the A4s was boosted even further on 27 August 1936 when the London-bound Silver Jubilee, hauled by No 2512 *Silver Fox*, managed 113 mph south of Grantham. That feat entered the record books as the fastest speed ever recorded by a revenue-earning train but, ironically, the eventual arrival at Kings Cross was seven minutes behind schedule because of the failure of the middle big-end near

TABLE 4.1: ORIGINAL DIMENSIONS OF THE A4 CLASS 4-6-2s

BUILT:	Doncaster 1935–1938
WEIGHTS FULL (locomotive):	102 tons 19 cwt
(tender):	60 tons 7 cwt/64 tons 3 cwt (a)
TENDER CAPACITY (water):	5,000 gallons
(coal):	8 tons/9 tons (b)
WHEELBASE (locomotive):	35ft 9in
(tender):	16ft 0in
WHEEL DIAMETERS (leading):	3ft 2in
(coupled):	6ft 8in
(trailing):	3ft 8in
CYLINDERS:	(3) $18\frac{1}{2}$in × 26in
BOILER PRESSURE:	250lb
HEATING SURFACES (tubes):	1281.4 sq ft
(flues):	1063.7 sq ft
(firebox):	231.2 sq ft
(superheater):	748.9 sq ft
SUPERHEATER:	Robinson 43-element
GRATE AREA:	41.25 sq ft
TRACTIVE EFFORT @ 85% boiler pressure:	35,455lb
LNER ROUTE AVAILABILITY:	9
BR POWER CLASSIFICATION:	7P (8P from May 1953)

Notes:

(a) The heavier weights apply to the corridor tenders.

(b) The smaller capacity applied to the corridor tenders. From 1937, their capacities were increased to 9 tons.

The famous Flying Scotsman service was soon treated to the attentions of the A4s. In this picture, No 2509 *Silver Link* is seen passing Low Fell with the southbound service in the autumn of 1935.

Photo: H. C. Casserley Collection

Hatfield. This was the first time an A4 had suffered major mechanical problems in service and the first total failure of an A4 did not occur until a week later, almost one year after the class had been introduced.

The moonlighting activities of the A4s on services other than the Silver Jubilee prompted the LNER to order seventeen more of the locomotives for general express work on the East Coast route. Numbered 4482–98, they were delivered between November 1936 and November 1937 and nine were treated to the conventional LNER green livery while the other eight wore Garter blue with red wheels. One of the duties which had been earmarked for the new arrivals was on the Coronation, a six-hour service between London and Edinburgh which was due to be inaugurated on 3 July 1937, with two stops on the northbound run and just one on the southbound journey. The original intention had been for the Coronation to operate as far as Aberdeen with three additional stops north of Edinburgh but, after much consideration, the LNER felt that it would make better business sense to

truncate the journey at Edinburgh. Passengers wishing to travel onwards to Aberdeen would be treated, as a bonus, to a trip behind a P2.

The theme which was originally selected for the names of the new locomotives was that of birds but, in May 1937, there was a hasty rethink. In recognition of the Imperial mood of the year, five of the new A4s were named, instead, after countries which belonged to the British Empire so that they would look more at home at the head of a train which bore the title of 'The Coronation'. When No 4489 was delivered in May of that year, it carried the name *Woodcock* in keeping with the original theme of bird names but, the following month, it was renamed *Dominion of Canada*. Ever conscious of the pennies, the LNER did not fancy wasting the nameplates of *Woodcock* and so they were reused on No 4493 which appeared in July.

One conspicuous exception to the name themes of the A4s involved No 4498 which was the one-hundredth Gresley 4-6-2 to be built. At a special ceremony held on 26 November 1937 at, of all places, Marylebone station, the locomotive was

In this photograph, the LNER blue livery of No 4492 *Dominion of New Zealand* looks little different to the silver livery of the first four A4s. However, the appearance of the locomotive's number on the front of the casing provides a clear differential as the 'silver' quartet carried only cab-side numbers until the introduction of the LNER's renumbering scheme of 1943. The date of the picture is 31 July 1937, the train is the 10.00am Flying Scotsman from Kings Cross and the location is Greenwood.
Photo: LCGB/Ken Nunn Collection

A4 No 4496 originally carried the name *Golden Shuttle* but was renamed *Dwight D. Eisenhower* in September 1945. Neither the location nor the date of this photograph have been recorded but it is quite possible that the locomotive was fresh from the works when it posed.

Photo: H. C. Casserley Collection

named *Sir Nigel Gresley* and this not only expressed the high regard in which the LNER's directors held their chief mechanical engineer but it also acknowledged the knighthood which Gresley had received in July the previous year for his contribution to railway engineering. At the naming ceremony, Gresley expressed his delight in seeing Oliver Bulleid who had been given a day off by his new bosses on the Southern Railway, where he had started work two months previously.

The Coronation was not the only A4-hauled named train to be introduced by the LNER in 1937 as, on 27 September that year, the West Riding Limited was inaugurated with a schedule of three hours and five minutes between Bradford and Kings Cross. The only intermediate stop was at Leeds and, from there, the eight-coach train was hauled by No 4492 *Dominion of New Zealand* but, as the A4s were too heavy for the first stretch from Bradford to Leeds, a pair of 0-6-2Ts had to be used for that first section of the journey. The appearance of No 4492 on the inaugural run had come about by default as the designated steed, No 4495 *Golden Fleece*, had broken down on the journey to Leeds the previous evening. It had been intended to reserve No 4495 and No 4496 *Golden Shuttle* for duties on the West Riding as far as possible and, for this purpose, they were the only A4s apart from No

4498 which carried non-standard names. Prior to its allocation to the West Riding service, No 4495 had been named *Great Snipe* and, avoiding wastage once again, the nameplates were saved and given to a later locomotive, No 4462.

The streamlined carriages of the Coronation and the West Riding Limited were painted in a blue livery and this looked rather swish as long as they were hauled by one of the blue locomotives. Similarly, the Silver Jubilee's rolling stock provided a natty match for the four silver-grey locomotives, but the LNER found it impossible to guarantee the use of the correctly coloured locomotives for any of the three trains. The great inconvenience of trying to run a railway by the dictates of colour schemes had to be dispensed with and, from October 1937, the livery of Garter blue with red wheels was adopted as standard for the entire class. The silver and the green locomotives were repainted when they entered the workshops for major services. There was one minor differential which was retained by the 'designated' locomotives, Nos 2509–12 and 4488–92/95/96, and this was the use of a red background on their nameplates as opposed to the standard black of their classmates.

Construction of more A4s continued from the end of 1937 and, by July 1938, a further fourteen had been delivered to complete the class with a total of thirty-five locomotives. The new arrivals, which took Nos 4462–69, 4499/4500 and 4900–03, were all named after birds and, as the A4s had, by this time, become regarded as all-purpose express locomotives, the new ones were paired with non-corridor tenders. It was one of these new A4s, No 4468 *Mallard*, which was to gain world-wide fame.

On 3 July 1938, driver Joseph Duddington and fireman Thomas Bray, both of Doncaster shed, were in charge of *Mallard* which had a dynamo-meter car and three twin-articulated Coronation coaches attached. The purpose of the special trip was, ostensibly, to carry out braking tests between Peterborough and London but Gresley made a great play of explaining that, in order to test the Westinghouse brakes to the limit, the train would have to be hauled at very high speeds before the

brakes were applied. The gaps between the lines in Gresley's statement were conspicuous, particularly as LMS No 6220 *Coronation* had notched up 114 mph only four days previously to snatch the British speed record for steam haulage and, one month before that, a German Federal Railways 4-6-4 had taken the world record with 124.5 mph.

Mallard's famous southbound run started at Grantham and, despite a slow start due to perm-anent way restrictions, the train was already doing 74.5 mph when it reached the summit of Stoke Bank. From that point, it was a downhill run and, for two miles between Little Bytham and Essendine, the train's speed did not drop below 120 mph. One and a half miles before Essendine, the peak of 126 mph was reached. Being the perfect gentleman, and a Knight of the Realm to boot, Gresley pointed out that, as the speed of 126 mph had been touched

TABLE 4.2: ORIGINAL NUMBERING AND NAMING OF A4 CLASS 4-6-2s

FIRST NO.	NAME	DATE BUILT	FIRST TENDER	FIRST SHED	1943 NO.	1946 NO.
2509	*Silver Link*	9/35	Corr	Kings Cross	580	14
2510	*Quicksilver*	9/35	Corr	Kings Cross	581	15
2511	*Silver King*	11/35	Corr	Gateshead	582	16
2512	*Silver Fox*	12/35	Corr	Kings Cross	583	17
4482	*Golden Eagle*	12/36	Corr	Kings Cross	584	23
4483	*Kingfisher*	12/36	Corr	Haymarket	585†	24
4484	*Falcon*	2/37	Corr	Haymarket	586	25
4485	*Kestrel*	2/37	Corr	Haymarket	587†	26
4486	*Merlin*	3/37	Corr	Haymarket	588†	27
4487	*Sea Eagle*	4/37	Corr	Gateshead	589	28
4488	*Union of South Africa*	6/37	Corr	Haymarket	590	9
4489	*Woodcock* (*)	5/37	Corr	Kings Cross	591	10
4490	*Empire of India*	6/37	Corr	Kings Cross	592	11
4491	*Commonwealth of Australia*	6/37	Corr	Haymarket	593	12
4492	*Dominion of New Zealand*	6/37	Corr	Kings Cross	594	13
4493	*Woodcock*	7/37	Corr	Gateshead	595	29
4494	*Osprey*	8/37	Corr	Heaton	596	3
4495	*Great Snipe* (*)	8/37	Corr	Doncaster	597	30
4496	*Golden Shuttle*	9/37	Corr	Doncaster	598	8
4497	*Golden Plover*	10/37	Corr	Haymarket	599	31
4498	*Sir Nigel Gresley*	11/37	Corr	Kings Cross	600	7
4462	*Great Snipe*	11/37	Non-C	Kings Cross	601	4
4463	*Sparrow Hawk*	12/37	Non-C	Gateshead	602	18
4464	*Bittern*	12/37	Non-C	Heaton	603	19

TABLE 4.2: CONTINUED

FIRST NO.	NAME	DATE BUILT	FIRST TENDER	FIRST SHED	1943 NO.	1946 NO.
4465	*Guillemot*	12/37	Non-C	Gateshead	604	20
4466	*Herring Gull*	1/38	Non-C	Kings Cross	605†	6
4467	*Wild Swan*	2/38	Non-C	Kings Cross	606	21
4468	*Mallard*	3/38	Non-C	Doncaster	607	22
4469	*Gadwall*	3/38	Non-C	Gateshead	(a)	(a)
4499	*Pochard*	4/38	Non-C	Gateshead	608	2
4500	*Garganey*	4/38	Non-C	Gateshead	609	1
4900	*Gannet*	5/38	Non-C	Doncaster	610	32
4901	*Capercaillie*	6/38	Non-C	Gateshead	611	5
4902	*Seagull*	7/38	Non-C	Kings Cross	612	33
4903	*Peregrine*	7/38	Non-C	Doncaster	613	34

Notes:

(*) Locomotives renamed and original names used for later A4s.

(†) Only these locomotives actually carried their allotted numbers.

(a) No 4469 withdrawn June 1942 after sustaining bomb damage.

The world-famous No 60022 *Mallard* is seen hauling the Capitals Limited near Durham on 16 July 1951. Its livery at the time was BR blue; the standard green livery was not applied until the following month.

Photo: E. R. Morten

only very briefly, the sustained speed of 125 mph would be the appropriate figure to be noted in the books. Even without that extra mile per hour, *Mallard's* achievement was enough to take the world record and, as it was never broken while steam still ruled, it is somewhat unlikely that it will ever be bettered in the future.

The only disappointment on that famous day was that, immediately after the record had been broken, *Mallard's* middle big-end bearing showed signs of overheating and, as a safety precaution, the locomotive was taken off the train at Peterborough. It was replaced by an ageing Ivatt 4-4-2 which hauled the train for the rest of the journey and, when it arrived at Kings Cross, it caused much confusion among the waiting pressmen. Although many journalists were unable to tell an Ivatt Atlantic from a Sopwith Camel, something seemed not quite right about the sight of the vintage ex-Great Northern engine at the head of the record-breaking train. A consequence of *Mallard's* retirement at Peterborough was that the alleged purpose of the trip, the brake tests, had to be abandoned but it was reported that Gresley was less distraught at this than the chaps from Westinghouse. Throughout the eventful day, there was one particular point that nobody had the heart to mention to Sir Nigel Gresley. This was the fact that the official speed limit on Stoke Bank was 90 mph.

The main reason why *Mallard* had been selected for the record-breaking run had nothing to do with Gresley's hobby of breeding wildfowl at his Hertfordshire home. The decision had been made purely for mechanical reasons, as the locomotive was the first of the class to have been built with the Kylchap double exhaust system. The last three A4s to be constructed, Nos 4901/02/03, were the only other members of the class to have double chimneys from new and, although all of the other A4s were to finish up with similar fitments, it was not until May 1957 when any were applied.

During 1939, the A4s became firmly established as the crack express locomotives in Britain and, apart from hauling the LNER's streamlined trains, they performed on most of the LNER's other express turns and also put in occasional guest appearances on lower-profile workings. Much to Gresley's immense satisfaction, the A4s had shown a marked

contrast to the W1 4-6-4 and P2 2-8-2s in that they had been conspicuously free from teething troubles and so the need for modifications had been virtually nil.

The demand for seats on A4-hauled streamlined train services was such that plans were formulated to introduce longer trains on the same routes. Despite the increased weights of the proposed trains, the LNER's customary high speeds would have to be maintained and so Gresley prepared the designs for a modified A4 with a boiler pressure of 275lb which, on paper, produced a tractive effort of 39,040lb as opposed to the existing 35,455lb. When the whispers about an 'improved' A4 became known outside the LNER, the engineering fraternity undertook debates, wrote magazine articles and even placed bets about the precise form which the new locomotive would take but, somewhere along the line, the old maxim about counting chickens had been forgotten. The real threat of war later in 1939 meant that the plans for an upgraded A4 were shelved and, eventually, abandoned.

The outbreak of World War Two completely changed the working patterns of the A4s. The LNER demonstrated its opinion of Neville Chamberlain's gullibility by suspending its three streamlined trains on 31 August 1939, three days before 'peace in our time' officially came to an end. At first, the eleven A4s which lived at Kings Cross shed were put into storage along with the streamlined coaches but when it was realised that, despite the initial optimism, the war was not going to be over by Christmas, the London-based A4s were returned to traffic for general duties. During the war years, the speed of the A4s was far less important than their haulage capacity, as passenger trains of more than twenty coaches became commonplace.

Under the demands of wartime traffic, the A4s performed frequent feats of both strength and speed as if to point out that the pre-war proposal for an improved version had been unnecessary. One legendary wartime run featured one of the double-chimney A4s, No 4901 *Capercaille*, which hauled a train of twenty-two bogie coaches, 730 tons in all, from Newcastle to Kings Cross and touched the speed of 78.5 mph in the process. The class leader, No 2509 *Silver Link*, was also credited with one particularly outstanding run which

A London-bound coal working on 12 June 1957 was a far cry from the inaugural duties on the Silver Jubilee which took pride of place in the CV of No 60014 *Silver Link*. The 34A shedplate can be seen quite clearly and, apart from two spells at Grantham, No 60014 was always a Kings Cross engine.

Photo: E. H. Sawford

involved hauling a twenty-five coach train of some 850 tons from Kings Cross to Newcastle and arriving within four minutes of schedule.

Understandably, the economies of the war years meant that the A4s suffered from the same lack of mechanical and cosmetic maintenance as all other locomotives. Right from the early days of wartime traffic conditions, the A4s proved that they were no posing prima donnas and, when it came to the application of unlined black liveries, they neither expected nor received any special treatment. During workshop visits, it was found that lower parts of the streamline casing hindered access and, as time was particularly valuable during the war, it was decided to remove the skirting below the running plate. The last A4 to have its lower valances removed was No 4462 *Great Snipe* which bared its moving bits to the outside world in October 1942 and, in common with all of its classmates, it was never to regain the skirting.

During an enemy air raid on the historic city of York on 29 April 1942, there was a direct hit on the locomotive depot and one of the A4s, No 4469, took much of the blast. The damage to the loco-

motive was so extensive that withdrawal was little more than a formality but, among the remnants salvaged from the wreck, were the bent but still legible nameplates of *Sir Ralph Wedgwood*. The wrecked A4 had been one of three locomotives which had been renamed after members of the LNER hierarchy in spring 1939. The gentlemen who had been thus honoured had witnessed the precedent set by the naming of No 4498 *Sir Nigel Gresley* in November 1937 and, much though they had respected their chief mechanical engineer, they had felt that, if a locomotive could be named after an employee, they should have a fair crack of the whip as well. Consequently, Nos 4469/99 and 4500 had their respective names of *Gadwall*, *Pochard* and *Garganey* replaced by *Sir Ralph Wedgwood*, *Sir Murrough Wilson* and *Sir Ronald Matthews*.

Sir Ralph Wedgwood was the chief general manager of the LNER and clearly thought more of prestige than superstition. After the demise of his customised A4, No 4469, during the York air raid, its name was saved and, in January 1944, was transferred to A4 No 4465 which had hitherto been named *Herring Gull*. During the strict economies of the war years, the LNER somehow saw fit to rename three of the A4s after directors of the company. Between July 1941 and October 1942, Nos 4462/94 and 4901 lost their respective names of *Great Snipe*, *Osprey* and *Capercaillie* in favour of *William Whitelaw*, *Andrew K. McCosh* and *Charles H. Newton*.

When No 4901's patron saint, Charles Newton, was knighted in 1943, little time was wasted in casting another set of new nameplates for the locomotive and, in June that year, it formally became *Sir Charles Newton*. The general request from the LNER's workforce was that, if Sir Charles were ever to receive letters after his name, could he please exercise modesty and wait until after the war before incurring the expense of yet more nameplates. A further renaming of an A4 took place immediately after the war when No 4496 had its nameplates of *Golden Shuttle* replaced by those of *Dwight D. Eisenhower*. Nobody on the LNER had the slightest wish to detract from Eisenhower's immense contribution to the Allied victory but, throughout the company, many comments were heard to the effect that it shouldn't have been too difficult to find the name of a suitable Briton instead.

The A4s were gradually restored to the impeccably-groomed conditions of their pre-war days but this came more out of prestige than necessity as, when the streamlined rolling stock of the Silver Jubilee and the West Riding train was reintroduced after the war, the carriages were designated for general, not specialised, usage. When No 4496 was renamed in September 1945, it was also treated to

the pre-war livery of Garter blue with red wheels, but it was June the following year before the next A4 was similarly treated. The last A4 to lose its black livery was, surprisingly, *Mallard* which did not regain the Garter blue finish until March 1948. Edward Thompson's renumbering scheme of 1943 allocated Nos 580–613 to the A4s in the order in which they had been constructed but, by the time the LNER renumbering of 1946 was announced, only four of the A4s had received their new numbers. These were Nos 4466/83/85/86 which had been given Nos 605/585/87/88 respectively.

Under the 1946 renumbering scheme, the A4s were designated Nos 1–34 but the order did not correspond with the previous sequence. Instead, the new system placed the locomotives into categories according to name themes and, quite predictably, the LNER's directors placed their own importance above all else. Numbers 1 to 6 were reserved for those A4s named after past and present company officials and, of course, the pecking order had to be respected. Therefore, No 1 went to the locomotive named after the current chairman, Sir Ronald Matthews while our old friend, Sir Charles Newton, entered the charts at No 5, having been pipped for the No 4 spot by ex-chairman William Whitelaw. The locomotive bearing the name of the famous designer of the celebrated class, *Sir Nigel Gresley*, had to take its turn and receive No 7, one in front of *Dwight D. Eisenhower*. Nos 9–13 went to the A4s bearing the names of countries of the Empire, Nos 14–17 were used by the 'Silver' engines while the 'bird' and 'Golden' names accounted for the rest.

TABLE 4.3: RENAMING OF A4s

DATE	NO.	NEW NAME	DATE	NO.	NEW NAME
6/37	4489	*Dominion of Canada*	10/42	4494	*Andrew K. McCosh*
9/37	4495	*Golden Fleece*	6/43	4901	*Sir Charles Newton*
3/39	4469	*Sir Ralph Wedgwood*	1/44	4466	*Sir Ralph Wedgwood*
3/39	4500	*Sir Ronald Matthews*	9/45	4496	*Dwight D. Eisenhower*
4/39	4499	*Sir Murrough Wilson*	10/47	4487*	*Walter K. Whigham*
7/41	4462	*William Whitelaw*	11/47	4485*	*Miles Beevor*
9/42	4901	*Charles H. Newton*			

* Nos 4485/87 carried Nos 26/28 at the time of being renamed.

Despite the lack of a headboard, the train is the northbound Coronation, known to all and sundry as the 4.00pm Kings Cross to Edinburgh. No 4498 *Sir Nigel Gresley* is the locomotive in charge as it passes near Potters Bar on 26 May 1938.
Photo: LCGB/Ken Nunn Collection

The directors' joint egotism made a mockery of the A4's status as one of the most famous locomotive classes in Britain. Sir Ronald Matthews might have been an exceptional chairman but, outside the railway world, his name meant very little, and so the naming of LNER No 1, the leader of the famous A4s, after him was never to register with the general public. In the eyes of all enthusiasts, there were three particular A4s which had far more genuine claims to the leadership. One was No 4468 *Mallard*, another was No 4498 *Sir Nigel Gresley* and the third was the original class leader, No 2509 *Silver Link*. Away from the boardroom, any of those three locomotive names meant far more than that of *Sir Ronald Matthews*.

Around the time of Nationalisation, a minor bout of renaming was inflicted on the A4s but this did not result in any renumbering as there was no intention to reflect social status. Late in 1947, Nos 26/28 (originally Nos 4485/87), had their respective names of *Kestrel* and *Sea Eagle* replaced by *Miles Beevor* and *Walter K. Whigham* while, in March 1948, BR No 60034 (originally LNER No 4903 and later No 34) lost its name of *Peregrine* to make way

for that of *Lord Faringdon*. Fortunately, the real-life Walter K. Whigham read nothing into his placement between *Merlin* and *Woodcock* and, similarly, the real Lord Faringdon had no strong objection to being positioned below *Seagull*.

Under the British Railways renumbering of 1948, the A4s took Nos 60001–34 and in June of that year, four of the class, Nos 60024/27/28/29, were painted in an experimental livery of purple. Despite later implications, hallucinogenic chemicals were not easily available in those days. In May 1949, the less garish British Railways express locomotive livery of ultramarine was applied to No 60013 and, by December 1950, all of the A4s sported this colour scheme. The blue was a darker shade than the LNER's Garter blue and was finished off with black and white lining. It was later decided to drop the blue livery in favour of dark green with orange and black lining and so, between August 1951 and November 1952, the A4s all underwent yet another change of paintwork. In those early years of British Railways, paintshop expenditure was rumoured to be the equivalent of the LNER's bill for constructing thirty-five A4s.

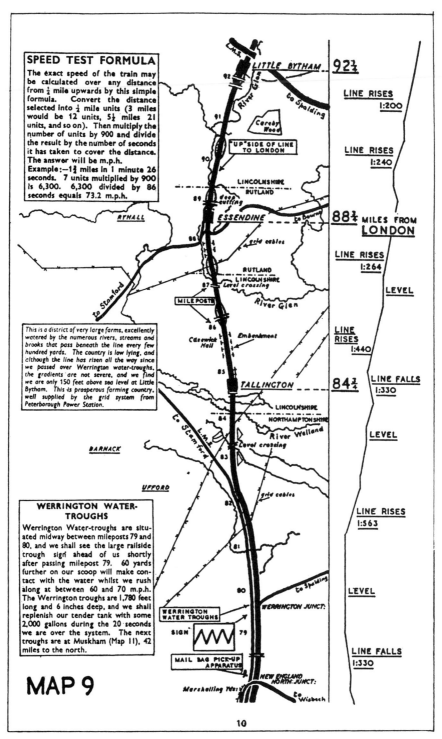

SPEED TEST FORMULA

The exact speed of the train may be calculated over any distance from ¼ mile upwards by this simple formula. Convert the distance selected into ¼ mile units (3 miles would be 12 units, 5¼ miles 21 units, and so on). Then multiply the number of units by 900 and divide the result by the number of seconds it has taken to cover the distance. The answer will be m.p.h.

Example:—1¾ miles in 1 minute 26 seconds. 7 units multiplied by 900 is 6,300. 6,300 divided by 86 seconds equals 73.2 m.p.h.

This is a district of very large farms, excellently watered by the numerous rivers, streams and brooks that pass beneath the line every few hundred yards. The country is low lying, and although the line has risen all the way since we passed over Werrington water-troughs, the gradients are not severe, and we find we are only 150 feet above sea level at Little Bytham. This is prosperous farming country, well supplied by the grid system from Peterborough Power Station.

WERRINGTON WATER-TROUGHS

Werrington Water-troughs are situated midway between mileposts 79 and 80, and we shall see the large railside trough sign ahead of us shortly after passing milepost 79. 60 yards further on our scoop will make contact with the water whilst we rush along at between 60 and 70 m.p.h. The Werrington troughs are 1,780 feet long and 6 inches deep, and we shall replenish our tender tank with some 2,000 gallons during the 20 seconds we are over the system. The next troughs are at Muskham (Map 11), 42 miles to the north.

MAP 9

10

The LNER took pride in its 'Mile by Mile' booklet which described the entire East Coast main line journey from Kings Cross through to Edinburgh. This page from the 1947 edition describes the delights of the line north of Peterborough but, remarkably, does not mention the feats of *Mallard* or *Papyrus* over this section.

TABLE 4.4: BR NUMBERING AND ALLOCATIONS OF A4 CLASS LOCOMOTIVES

NO.	NAME	1/1/48 SHED	1/1/58 SHED	LAST SHED	WDN.
60001	*Sir Ronald Matthews*	Gateshead	Gateshead	Gateshead	10/64
60002	*Sir Murrough Wilson*	Gateshead	Gateshead	Gateshead	5/64
60003	*Andrew K. McCosh*	Kings Cross	Kings Cross	Kings Cross	12/62
60004	*William Whitelaw*	Haymarket	Haymarket	Ferryhill	7/66
60005	*Sir Charles Newton*	Gateshead	Gateshead	Ferryhill	3/64
60006	*Sir Ralph Wedgwood*	Kings Cross	Kings Cross	Ferryhill	9/65
60007	*Sir Nigel Gresley*	Grantham	Kings Cross	Ferryhill	2/66
60008	*Dwight D. Eisenhower*	Grantham	Kings Cross	New England	7/63
60009	*Union of South Africa*	Haymarket	Haymarket	Ferryhill	6/66
60010	*Dominion of Canada*	Kings Cross	Kings Cross	Ferryhill	5/65
60011	*Empire of India*	Haymarket	Haymarket	Ferryhill	5/64
60012	*Commonwealth of Australia*	Haymarket	Haymarket	Ferryhill	8/64
60013	*Dominion of New Zealand*	Kings Cross	Kings Cross	Kings Cross	4/63
60014	*Silver Link*	Grantham	Kings Cross	Kings Cross	12/62
60015	*Quicksilver*	Grantham	Kings Cross	Kings Cross	4/63
60016	*Silver King*	Gateshead	Gateshead	Ferryhill	3/65
60017	*Silver Fox*	Kings Cross	Kings Cross	New England	10/63
60018	*Sparrow Hawk*	Gateshead	Gateshead	Gateshead	6/63
60019	*Bittern*	Gateshead	Gateshead	Ferryhill	9/66
60020	*Guillemot*	Gateshead	Gateshead	Gateshead	3/64
60021	*Wild Swan*	Kings Cross	Kings Cross	New England	10/63
60022	*Mallard*	Grantham	Kings Cross	Kings Cross	4/63
60023	*Golden Eagle*	Gateshead	Gateshead	Ferryhill	10/64
60024	*Kingfisher*	Haymarket	Haymarket	Ferryhill	9/66
60025	*Falcon*	Kings Cross	Kings Cross	New England	10/63
60026	*Miles Beevor*	Kings Cross	Kings Cross	Ferryhill	12/65
60027	*Merlin*	Haymarket	Haymarket	St. Margarets	9/65
60028	*Walter K. Whigham*	Grantham	Kings Cross	Kings Cross	12/62
60029	*Woodcock*	Kings Cross	Kings Cross	New England	10/63
60030	*Golden Fleece*	Grantham	Kings Cross	Kings Cross	12/62
60031	*Golden Plover*	Haymarket	Haymarket	St. Rollox	10/65
60032	*Gannet*	Grantham	Kings Cross	New England	10/63
60033	*Seagull*	Grantham	Kings Cross	Kings Cross	12/62
60034	* *Lord Faringdon*	Grantham	Kings Cross	Ferryhill	8/66

* No 60034 named *Peregrine* until March 1948.

During April and May 1948, British Railways initiated interchange trials with express locomotives from each of its newly-formed regions. The purpose was not to pit one type against another but to gather information for any future standardisation of motive power. The Midland Region nominated Coronation class 4-6-2s, the Western offered King class 4-6-0s and the Southern provided Merchant Navy 4-6-2s which had been designed by Gresley's former right-hand man, Oliver Bullied. The Eastern Region, quite naturally, selected A4s.

On the Eastern Region, the proving ground was the run between Kings Cross and Leeds and, as the crew of No 60034 *Lord Faringdon* were not exactly

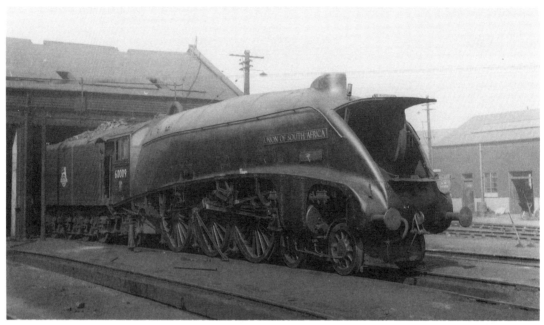

On a cold winter's day in January 1956, the smokebox of No 60009 *Union of South Africa* was probably a sought after source of warmth for the cleaners at Haymarket shed.

Photo: D. K. Jones Collection

unfamiliar with the line, it was unsurprising that the A4 performed more economically than its rivals. The Midland Region's test route was between Euston and Carlisle and, on the tricky haul over Shap, the best run was provided, once again, by No 60034. It was the Southern Region which had the celebrity treatment as, for the tests between Waterloo and Exeter, the LNER's representative was No 60022 *Mallard* while, on the Western Region's route between Paddington and Plymouth, No 60033 *Seagull* was used. On both the Southern and the Western Regions, the A4s performed superbly.

The results from all of the tests proved conclusively that the A4s were more efficient and more economical locomotives than any of the others. This did not only provide the ex-LNER men with the satisfaction of reversing the embarrassment of the 1925 trials when a GWR Castle had outstripped an A1. It also provided realistic expectations that the lessons learned would result in any future designs for standard British Railways express locomotives leaning on the expertise of Doncaster. As if to demonstrate his ability to absorb and assess factual information, Robert Riddles went not to Darlington

but to Derby to have his Britannia 4-6-2s designed and to Crewe to have them built.

At the time of Nationalisation, the allocation of the A4s was divided among Kings Cross, Grantham, Gateshead and Haymarket. The Haymarket contingent of seven had rarely worked beyond Newcastle since the war but, when the non-stop Flying Scotsman was reintroduced on 21 May 1948, they shared the duty with their counterparts at Kings Cross. The normal Flying Scotsman service was rudely interrupted on 12 August when storms caused extensive damage to the line between Berwick and Dunbar but, rather than suspend the service, it was decided to reroute it via Galashiels. Over this line, the stipulated maximum loading for a 4-6-2 was 400 tons and, with gradients of up to 1 in 70 on the southbound run and a shortage of water troughs, it was never anticipated that the rerouted train would operate non-stop. The Haymarket crews, however, always relished a challenge and, on 24 August 1948, No 60029 *Woodcock* left Edinburgh at the head of a 435-ton Flying Scotsman and it was taken through to Kings Cross without a break. The non-stop run of 408½ miles was a world record

Table 1 LONDON (King's Cross), PETERBOROUGH, GRANTHAM, DONCASTER, LEEDS, BRADFORD, HULL, & YORK

MONDAYS TO FRIDAYS

Table 1

Mondays to Fridays   Mondays to Fridays

Table 1

[A large, dense railway timetable grid follows, listing stations down the left column with mileage, and numerous columns of departure/arrival times. Named expresses indicated in vertical text include: THE WHITE ROSE, THE CAPITALS LIMITED (Non-stop King's Cross to Edinburgh), THE NORSEMAN, THE FLYING SCOTSMAN. Stations listed include:]

- LONDON King's Cross (dep)
- Finsbury Park
- Hatfield
- CAMBRIDGE (dep)
- Hitchin (dep)
- Three Counties
- Arlesey and Henlow
- Biggleswade
- Sandy
- Tempsford
- St. Neots
- Offord and Buckden
- CAMBRIDGE (arr)
- Huntingdon (North)
- Abbots Ripton
- Holme
- Yaxley and Farcet
- Peterboro' (North) (arr)
- BOSTON
- GRIMSBY Town
- CROMER (Beach)
- NEWMARKET
- CAMBRIDGE
- HARWICH Town
- IPSWICH
- YARMOUTH (V.)
- LOWESTOFT
- CROMER (High)
- NORWICH (Thorpe)
- Peterboro' (North) (dep)
- Tallington
- Essendine
- Little Bytham
- Corby Glen
- Great Ponton
- Grantham (arr)
- LINCOLN (Central)
- Grantham (dep)
- Barkston
- Hougham
- Claypole
- Newark (North Gate)
- Carlton-on-Trent
- Crow Park for Sutton-on-T.
- Tuxford (North)
- Retford
- LINCOLN (Central) (arr)
- GRIMSBY Town
- SHEFFIELD (Victoria)
- MANCHESTER (L.Rd.)
- (Central)
- LIVERPOOL (Central)
- LINCOLN (Central) (dep)
- Retford (dep)
- Ranskill
- Bawtry
- Rossington
- LINCOLN (Central) (dep)
- Doncaster (arr)
- HULL (arr)
- Doncaster (dep)
- Wakefield (W'gate)
- Wakefield (Kirkgate)
- HUDDERSFIELD
- BRADFORD (Exch.)
- Goose Hill (High Level)
- HALIFAX Town
- HARROGATE
- Leeds (Central)
- Doncaster (dep)
- Arksey
- Moss
- Balne
- Hock
- Temple Hirst
- Selby (arr)
- HULL
- Selby (dep)
- Riccall
- Escrick
- Naburn
- York
- HARROGATE (arr)
- RIPON
- SCARBOROUGH (C.)
- DARLINGTON
- MIDDLESBROUGH
- SUNDERLAND
- NEWCASTLE (Wav.)
- EDINBURGH (Wav.)
- GLASGOW (Queen St.)
- FORT WILLIAM
- STIRLING
- PERTH (General)
- INVERNESS
- DUNDEE (Tay Bridge)
- ABERDEEN

Footnotes / reference notes:

a. Arr. 11 51 a.m. on Mondays and Fridays
a. a.m.
b. Via Leeds (Central)
b. Via Bradford
D. Passengers cross from Wakefield (Westgate) to (Kirkgate) at own expense
d. Via Leeds (Central) & (City)
d. Dep. Corby Glen 10 47 and arr. Grantham 11 2 a.m. on Wednesdays
e. Via Cambridge. Does not run on Thursdays from 24th July to 28th August incl.
e. Via Cambridge 9 5 p.m. on Sundays

F. Dep. Cambridge 9 46, Yarmouth (V.) 8 40, Lowestoft (Central) 9 10, Cromer (High) 8 40, and Norwich (Thorpe) 10 5 p.m. on Sundays
f. Arr. Boston 11 12 a.m. on Thursdays from 24th July to 28th August inclusive
g. Via Grantham. Does not run on Thursdays from 24th July to 28th August incl.
H. Via Holbeck
h. Via Retford and Penistone Applies only to passengers from beyond Retford

d. Arr. 3 minutes earlier
J. On Wednesdays from 9th July to 3rd Sept. dep. Yarmouth (V.) 10 0 p.m. Dep. Cambridge 1 8 a.m. on Mondays, Harwich Town 9 15, Yarmouth (V.) 10 45, Lowestoft (C.) 9 10, and Cromer (High) 8 40 p.m. on Sundays
k. Arr. 3 48 p.m. on Fridays 18th July to 5th September inclusive
k. Change at Welwyn Garden City

M. Passengers change at Huntingdon from East to North (distance 100 yards)
n. Change at Hitchin
n. Via Grantham. On Thursdays from 24th July to 28th August inclusive arr. Boston 11 12 a.m. via Spalding
o. Newcastle (Tyne Commission Quay)
p. p.m.
R. For Sheffield, Manchester, and Liverpool
RC. Refreshment Car
SC. Limited Sleeping accommodation

t. Norwich (City) via Sutton Bridge
TC. Through Carriages
k. Beach Station via Sutton Bridge
x. Arr. 4 minutes earlier
Z. Arr. Darlington 10 7, Newcastle 10 42 a.m., Edinburgh 1 29, and Glasgow 2 49 p.m. on Mondays
*. Passengers cross from Manchester (Lon. Rd.) to (Cen.) at own expense
§. Arr. 11 5 a.m. Mondays

This is just a single page of the Eastern Region's summer 1952 timetable for the East Coast main line but four named expresses are shown to leave Kings Cross within one hour.

and the A4s were to repeat the feat another sixteen times before the main line was reopened.

During 1950 and 1951, an extensive revision of working arrangements resulted in all of Grantham's A4s being transferred to Kings Cross and so the task of inaugurating new northbound services inevitably fell to the London-based locomotives. One of these was the Capitals Limited, a non-stop service between Kings Cross and Edinburgh, which was introduced for the summer of 1951 with a schedule of seven hours and twenty minutes. The A4s were to prove well in control of things on the Capitals and, for the summer of 1952, the timing for the train was safely reduced to seven hours and six minutes. For the summer of 1953, the Capitals was renamed the Elizabethan to commemorate the change of tenancy at Head Office, London SW1 and the first down

journey with the new headboard was given to No 60028 *Walter K. Whigham* while No 60009 *Union of South Africa* performed the honours in the southbound direction.

During the early 1950s, the working of three Pullman trains, the Queen of Scots, the Yorkshire and the Tyne-Tees, usually fell to the A4s and later, in September 1956, the Talisman was introduced with a six and a half-hour schedule between Kings Cross and Edinburgh. The only stop was at Newcastle where the engines were changed and, on the inaugural services, Kings Cross-based No 60025 *Falcon* took the northbound train while No 60019 *Bittern* of Gateshead hauled the southbound train from Newcastle. It was rare for the A4s to stray from the East Coast main line and unusual workings were usually accounted for by the locomotives'

A smart-looking No 60009 *Union of South Africa* is seen backing out of Kings Cross in September 1961 after working the southbound Elizabethan. Apart from No 60001 and the prematurely-retired LNER No 4469 which both remained allocated to Gateshead depot throughout their entire lives, No 60009 was one of the most home-loving of the A4s. It only ever had one allocation change and that was in May 1962 when it was transferred from Haymarket to Aberdeen. The only other A4s which were subjected to just one transfer throughout their existences were No 60017 and No 60031.

Photo: E. H. Sawford

popularity for specials. Any other spells abroad, such as workings between Glasgow and Carlisle or over the Waverley line, were usually brief.

Throughout the 1950s, the A4s continued to perform excellent work and, despite official speed restrictions, the occasional ton-up proved no real obstacle to the locomotives. Occasionally, however, there were reports of poor steaming from an A4 and, although it was usually the same few locomotives which featured in most of the reports, various experiments with blastpipes were undertaken. On the whole, the tests proved inconclusive as the steaming problems seemed only to manifest themselves when poor quality coal was being used but, nevertheless, some improvements to the blastpipes were eventually obtained. Early in 1957, somebody saw fit to remark that, of the A4s which had ever been accused of poor steaming or troublesome blastpipes, the list had *not* included Nos 60005/22/33/34, the four which had been fitted with double chimneys and blastpipes from new. The fitting of double chimneys and Kylchap blastpipes to the other thirty A4s was undertaken between May 1957 and November 1958 and virtually cured the problems overnight.

Diesel traction was introduced with a vengeance on the East Coast main line at the end of 1958 in the form of English Electric Type 4s but, much to the delight of steam enthusiasts, the high failure rate of the Type 4s meant that a number of trains scheduled for diesel haulage turned up, instead, with an A4 at the helm. By contrast, the scheduled arrival of the Deltic diesels for 1961 was a far more serious affair for the A4s as the internal-combustion creatures were intended for heavy express work on the East Coast main line. The Deltics did not, however, materialise quite as quickly as anticipated and so, when three special trains were required to run from Kings Cross to York for the wedding of the Duke of Kent on 8 June 1961, the duties were allocated to A4s. No 60003 *Andrew K. McCosh*, No 60015 *Quicksilver* and No 60028 *Walter K. Whigham* were the locomotives involved, and No 60014 *Silver Link* was retained as a standby.

When the Deltics made their belated appearances, the first casualty for the A4s was the loss of their monopoly of the Elizabethan. These turns were gradually taken over by the Deltics during the summer of 1961 although the final use of steam haulage of the train was not until 9 September 1961, just one day before the service was suspended for the winter. The locomotives for that last steam trip were No 60022 *Mallard* on the northbound trip and No 60009 *Union of South Africa* on the southbound. During that same summer, duties on the Queen of Scots Pullman were taken over completely by diesels, but in this case, the invaders were the Type 4s. The final non-stop steam working between Kings Cross and Edinburgh was an RCTS special which was hauled by *Mallard* on 2 March 1962. The train continued from Edinburgh to Aberdeen behind No 60004 *William Whitelaw*.

For many enthusiasts, it seemed impossible to believe that one of Britain's most famous classes of locomotives would actually be dispensed with. One of the very few classes which could be compared to the A4s was the ex-Great Western King class 4-6-0s and the shock waves which accompanied the unceremonious withdrawal of all thirty Kings during the summer and autumn of 1962 acted as a reminder to all A4 enthusiasts that, in railway terms, nothing was sacred. Reality set in on 29 December that same year when Nos 60003/14/28/30/33 were withdrawn simultaneously from Kings Cross shed. The closure of the shed in June 1963 resulted in the eleven remaining A4s being transferred to New England shed at Peterborough. Little over four months later, on 29 October 1963, the departure of No 60017 *Silver Fox* on the 6.40pm to Leeds became the last appearance of an A4 on a scheduled service at Kings Cross.

Withdrawal of the A4s continued throughout 1963 and many of those which escaped were despatched to St. Margaret's shed in Edinburgh although, later, they were concentrated at Ferryhill depot in Aberdeen.

In Scotland, they were treated with due respect and, instead of being relegated to secondary or mixed traffic duties, they were regularly used on fast passenger services between Glasgow and Aberdeen via Perth and, occasionally, between Edinburgh and Aberdeen. The reprieve could not, of course, last forever. One by one, they were retired and, by the beginning of 1966, only six A4s remained in service. During the first week of September that year, the last two survivors, No

60019 *Bittern* and No 60024 *Kingfisher*, were withdrawn from their final home at Ferryhill.

That may have been the end for the A4s on British Railways but six of the class passed into preservation and, of those, four are alive and well and working in Britain. It would have taken a very brave individual to order the cutting up of No 60022 *Mallard* and so, when it was retired from Kings Cross in April 1963, there was little doubt that it would get the preservation treatment. A superb restoration job was carried out at Doncaster works and the locomotive was returned to its original condition, complete with Garter blue livery and the number 4468. It would take an exceptionally pernickety enthusiast to quibble about the fact that the tender with which it is now paired last belonged to No 60026 *Miles Beevor* and not *Mallard* itself.

Two months after the withdrawal of *Mallard*, No 60008 *Dwight D. Eisenhower* was retired and despatched to Doncaster works. Unlike *Mallard*, it was not restored to its original condition but, instead, given a full cosmetic overhaul in its British Railways retirement outfit in preparation for its new home at the National Railroad Museum in Wisconsin, USA. The handing-over ceremony took place on 27 April 1964 when No 60008 was ready to be loaded on board a ship at Southampton Docks, and the representative of the British Railways Board on the day was one Dr. Richard Beeching. It must have been interesting to hear how he praised the design and the capabilities of the locomotive so soon after he had signed its death warrant.

The shedmaster at Aberdeen sent No 60010 *Dominion of Canada* to Darlington works for repair in May 1965 but, due to the condition of its boiler, he didn't get the engine back. After being hastily condemned, it was left to rust in peace at Bank Top shed until August 1966 when it was towed to Crewe works for a complete external overhaul. The insult of having the work done at Crewe was, unfortunately, unavoidable as Doncaster works had closed its doors to steam at the end of 1963. After leaving the works with a gleaming British Railways livery, No 60010 was shipped to Canada on 10 April 1967 for permanent static exhibition at the museum of the Canadian Railroad Historical Association near Montreal.

The first A4 to be privately preserved was No 60007 *Sir Nigel Gresley*. After its withdrawal from Aberdeen shed in February 1966, it was purchased by the A4 Locomotive Society and, although it was restored to the Garter blue livery with the number 4498, the Society decided not to go the whole hog and give it back its original single chimney and side skirting. Part of the reason for the compromise over the degree of restoration was that the Society was keen to woo British Railways in the hope that the locomotive would be cleared for use on main lines. The all-clear was given and, in March 1967, the locomotive was steamed once again in preparation for its return to the open road.

When No 60009 *Union of South Africa* was withdrawn from Aberdeen on 1 June 1966, it was still in a good general condition and so it was an obvious target for preservationists. The organisation which secured the locomotive was the Lochty Private Railway Company which laid its own three-mile line near Anstruther in Fife and, on 14 June 1967, the locomotive went into action for its new owners. It has since made appearances on other preserved lines in Britain. The sixth and final A4 to pass into preservation had, like the previous three, seen out its last days at Aberdeen. This was No

TABLE 4.5: YEARLY TOTALS OF A4 CLASS LOCOMOTIVES

Totals taken at 31 December each year.

1935	4	1940	35	1945	34	1950	34	1955	34	1959	34	1963	19
1936	6	1941	35	1946	34	1951	34	1956	34	1960	34	1964	12
1937	25	1942	34	1947	34	1952	34	1957	34	1961	34	1965	6
1938	35	1943	34	1948	34	1953	34	1958	34	1962	29	1966	0
1939	35	1944	34	1949	34	1954	34						

During the last years of the steam age, there were many sorry sights and, almost certainly, we all became somewhat immune to seeing our favourite locomotives looking totally neglected. Nevertheless, the sight of a string of diabolically filthy A4s must have shocked even the most hardened fatalist. At New England shed on 1 December 1963, the line-up of stored A4s comprised No 60017 *Silver Fox*, No 60029 *Woodcock*, No 60032 *Gannet*, No 60025 *Falcon* and No 60021 *Wild Swan*. Tucked amongst them is B1 4-6-0 No 61122. It seems that the only clean bits on No 60017 are its famous emblem and the 'overhead wires' plaque.

Photo: M. John Stretton

99

60019 *Bittern* which was one of the last two A4s to be withdrawn in September 1966 and, like No 60009, no significant repair work was required before it was put to work by its new owner, Mr. G. Drury. Between 1967 and 1972, No 60019 was one of many preserved locomotives which were sidelined by British Railways' controversial ban on main-line steam and, during that period, it was restored to a Garter blue livery with the post-1946 number of 19.

In recent years, both No 60009 and No 60019 have been given new identities. The former was renamed *Osprey*, the name which was originally designated for it in 1937 before the decision was made to call it, instead, *Union of South Africa*. The latter, No 60019, has been given No 2509 and renamed *Silver Link* in memory of the very first A4

although the real *Silver Link* was, quite inexplicably, cut up at Darlington in 1963. The North Eastern Locomotive Preservation Group's transformation of *Bittern* to *Silver Link* included full streamlined skirting and a silver-grey livery; a truly excellent job has been done and the locomotive is now the prized exhibit at the Stephenson Railway Museum on Tyneside.

Despite the best efforts of the British Railways Board, which let two of the class go abroad, the contingent of four active A4s in Britain is highly commendable. It is a tribute not only to the enthusiasm and the dedication of the preservationists, but also to the design of Nigel Gresley's masterpieces that, at the end of almost thirty years of high mileage express work, the A4s were raring to go when they were returned to the tracks.

CHAPTER 5

More Casings and Skirts

The A4 4-6-2s were, unquestionably, Nigel Gresley's masterpieces. The record-breaking exploits of No 4468 *Mallard* on 3 July 1938 provided the perfect illustration of their speed capabilities but they had, in fact, caught the public's imagination right from the time No 2509 *Silver Link* first poked its nose out of the doors of Doncaster works in September 1935. Much of the A4s' appeal was down to their revol-

utionary appearances, but they were not the only Gresley locomotives to receive the treatment of streamlining.

The W1 4-6-4 No 10000, which had been completed late in 1929, certainly fell into the streamlined category, but the external appearance of the locomotive had been largely dictated by the use of a water-tube boiler. When the 4-6-4 entered service,

The first two P2s, Nos 2001/02, had appeared in 1934 and the design of their front ends leant on that of the water-tube W1 No 10000. However, when four more P2s were ordered in 1936, the new quartet had their front ends fully enclosed in streamline casing similar to that of the A4 Pacifics. As the A4s were, at that time, the flavour of the month in the locomotive world, the two original P2s quickly cashed in on the fame of the Pacifics by having full front-end casing as well. The first of the 1936 P2s was No 2003 *Lord President* and its impressive appearance and awesome proportions can be clearly seen in this excellent picture.

Photo: Rail Archive Stephenson

its general behaviour and reliability proved a little less than awesome and, consequently, much of the attention it received was not directed at its imposing looks. The story of the W1 was echoed but, mercifully for Gresley, on a much reduced scale when the P2 2-8-2s, Nos 2001/02, took to the road in 1934. Externally, the pair of P2s borrowed a fair bit from the lines of the W1 and the locomotives justified the description of semi-streamlined but, in service, they did not prove as wonderful on the Edinburgh to Aberdeen route as had been anticipated. Any embarrassment that Gresley might have felt over the W1 and P2s disappeared overnight when the A4s went into service and, from autumn 1935, he could do little wrong.

Although the pair of P2s had presented a number of problems on their duties between Edinburgh and Aberdeen, the difficult nature of the route had defied attempts to come up with anything better. Therefore, when the subject of additional motive power for that line cropped up in 1936, it was decided to build four more P2s. The first of these was completed in June of the same year and, by September, the last was delivered. Mechanically, the new P2s were almost identical to their predecessors and perpetuated the combination of 6ft 2in coupled wheels, 21in × 26in cylinders and a 220lb boiler but, externally, they differed from the previous pair in having streamlined front-ends almost identical to those of the A4s. The new P2s continued the numbering sequence started by their predecessors and took Nos 2003–06; they were named *Lord President, Mons Meg, Thane of Fife* and *Wolf of Badenoch*. A former North-British Railway 4-4-2, latterly LNER No 9871, had previously carried the name *Thane of Fife* and it had been a classmate of that locomotive which had yielded the name of *Cock O' The North* to the original P2, No 2001.

Ostensibly, the purpose of the streamline casing around the front-ends was to help with smoke deflection, but the integral deflectors which had been fitted to the first two P2s had done a satisfactory, if not earth-shattering job. Although the P2s were powerful locomotives, they were required as much for strength as speed and, as the route to Aberdeen was not the place for potential boy racers, the streamlining could hardly be justified on the grounds of hair-raising schedules. The application of the casing to the front-ends of the new P2s was, in effect, little more than a cosmetic exercise which cashed in on the fame of the A4s but, nevertheless, it had the effect of making good-looking locomotives even more eye-catching.

Between Edinburgh and Dundee, the P2s were allowed loadings of 550 tons while, on the more difficult section between Dundee and Aberdeen, the limit was reduced to 530 tons. The operational difficulties which had been encountered with the first two P2s did not miraculously disappear when the new quartet arrived but, in view of the overall strength of the locomotives and the lack of any alternatives, the problems came to be tolerated. In October 1936, No 2002 *Earl Marischal* was rebuilt with a streamlined front-end similar to Nos 2003–06 and, in December 1937, No 2001 *Cock O' The North* followed suit. While No 2001 was in the workshop for its nose job, it was fitted with Walschaerts valve gear instead of the original Lentz equipment and its feed-water heater was removed.

TABLE 5.1: SUMMARY OF P2 CLASS 2-8-2s

N.B. Original dimensions listed in Chapter Three.

NO.	NAME	BUILT	WORKS NO.	REBUILT AS A2/2	BR NO.	WDN.
2001	*Cock O' The North*	5/34	1789	9/44	60501	2/60
2002	*Earl Marischal*	10/34	1796	6/44	60502	7/61
2003	*Lord President*	6/36	1836	12/44	60503	11/59
2004	*Mons Meg*	7/36	1839	11/44	60504	1/61
2005	*Thane of Fife*	8/36	1840	1/43	60505	11/59
2006	*Wolf of Badenoch*	9/36	1842	5/44	60506	4/61

The 4.00pm Kings Cross to Leeds gets under way on 21 August 1936 behind new P2 No 2004 *Mons Meg*. In the distance, a fair old crowd can be seen on the departure platform and considering that the locals had, by then, become well used to the A4s, the interest in the P2 speaks volumes for its cosmetic appeal. Either that or there was a fire drill.

Photo: LCGB/Ken Nunn Collection

With steam cocks open, P2 2-8-2 No 2004 *Mons Meg* is being prepared at Kings Cross station on 21 August 1936.

Photo: LCGB/Ken Nunn Collection

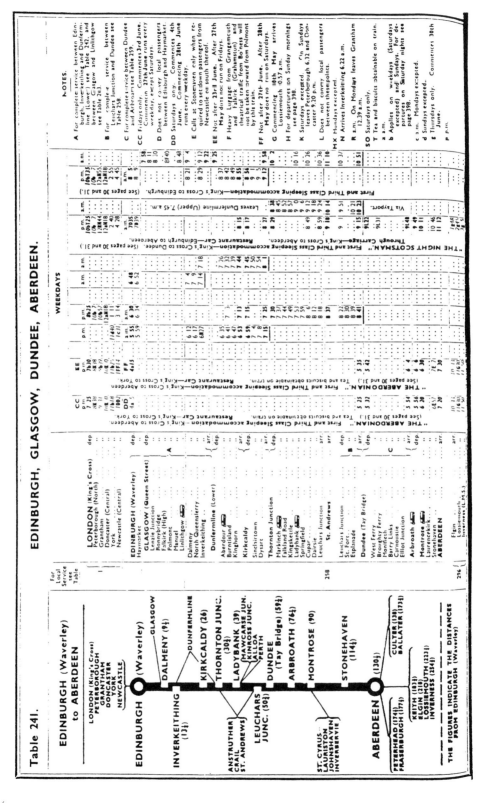

The difficult line between Edinburgh and Aberdeen provided the reason for the development of the P2 2-8-2s. This extract from the 1938 LNER timetable shows the services which would have been entrusted to those locomotives.

As a result of the modifications carried out to Nos 2001/02, the six P2s now looked more like a single class with overall similarities. Not only were their performances more consistent with each other but their nagging faults also displayed consistency. The overheating of the axleboxes and the failure of crank axles were put down to the stresses of the curving route on the locomotives' long wheelbases but it took some time to appreciate that the design of the swing-link pony truck was also a major cause of stress. The swing-link design had long since been discontinued for bogies, but the possibility of the same design causing problems for a single axle had hardly been considered.

With the outbreak of war in 1939, Gresley had no opportunity to experiment with alternative designs for the pony trucks of the P2s, partly because of a lack of resources but also because, on the Aberdeen run, an erratic P2 was far better than no P2 at all. When Edward Thompson succeeded Gresley in 1941, he grasped the opportunity to show what he thought of the P2s and swiftly earmarked them for conversion to 4-6-2s. Thompson had taken over the job as Gresley's principal assistant after Oliver Bulleid had joined the Southern Railway in 1937 and, whereas Gresley and Bulleid had had an excellent working relationship, the combination of Gresley and Thompson was not made in heaven.

Thompson's roots were with the North Eastern Railway and Darlington works, as opposed to Gresley's upbringing with the Great Northern at Doncaster. The last chief mechanical engineer of the North Eastern before the grouping was Sir Vincent Raven and, as he was Thompson's father-in-law, a division of loyalties was, to an extent, understandable. Along with several contemporaries, Thompson viewed Gresley's enthusiasm for Doncaster as an attempt to dispense with any North Eastern influence on the LNER and the rivalry between Darlington and Doncaster became far more than just matters of pride or tradition. Working for the LNER from his base at Darlington and, later, at Stratford, Thompson had to watch Gresley and Bulleid bask in the limelight time and time again. This did nothing to improve Thompson's frame of mind.

When Thompson succeeded Oliver Bulleid, Gresley might not have gone overboard about the personality or temperament of his new assistant but he had no doubts about Thompson's engineering capabilities. Thompson had often expressed his disagreement with certain Gresley policies but, as insecurity was not a part of Gresley's make-up, the voicing of different opinions was not seen as a threat. Thompson was a firm supporter of a two-cylinder arrangement instead of Gresley's favoured three-cylinders and he was also a keen advocate of standardisation. Whereas Gresley was once quoted as saying that 'to standardise is to stagnate', Thompson did not hold with the idea of developing what he considered to be extravagant designs for specialised tasks.

Thompson did not hesitate to place Gresley's P2s in his category of 'extravagances' and the first drawings for the conversion to 4-6-2s were made in April 1942, less than a year after he had succeeded Gresley. The plans included the removal of the streamline casing, and the first to succumb to total rebuilding was No 2005 *Thane of Fife* which re-emerged from the works on 16 January 1943. After extensive trials of No 2005, Nos 2001–04/06 were similarly treated between May and December 1944. With the new classification of A2/2, they were returned to their old haunts of Aberdeen and Edinburgh, which was probably just as well considering that they retained their original names. While *Cock O' The North* might have been a wonderfully patriotic name in Scotland, a locomotive bearing that name might have caused a few raised eyebrows among more genteel passengers at Kings Cross.

Prior to 1937, the application of streamlining had been reserved for locomotives at the top end of the power or size scales but, in autumn that year, a pair of B17 4-6-0s received the full streamline treatment. The purpose of this exercise was purely cosmetic, as it was planned to coincide with the introduction of a new named train, the East Anglian, between Liverpool Street and Norwich. A combination of the popularity of art deco and the reputation of Gresley's A4s was behind the decision to use streamlined locomotives on the new service, although some saw it as a sop to the staff and passengers on the former Great Eastern section. While the A4s and P2s had attracted a good deal of publicity elsewhere on the LNER, the weight restrictions of many of the lines in East Anglia rendered unlikely the possibility of those locomotives appearing at Liverpool Street. The provision of streamlined locomotives for a Liverpool Street service helped to prevent the natives from feeling left out.

The two B17s which were selected for new overcoats were No 2859 *Norwich City* and No 2870 *Tottenham Hotspur* and they were renamed *East Anglian* and *City of London* respectively. A pair of six-coach sets, each weighing 219 tons empty, was constructed at York specially for the new service but, while having similar luxurious internal fitments to the streamlined coaches of the West Riding and Silver Jubilee, there was no exterior streamlining. The service was inaugurated on 27 September 1937 but there was never any possibility of making it a true high-speed run. The density of traffic on the route prevented a schedule faster than two hours and eleven minutes for the 116 miles but, as if not to be overshadowed by their more celebrated streamlined counterparts, the two B17s were known, on occasions, to knock fifteen minutes off the schedule.

Even with a smart apple green livery, the effect of streamlining the B17s didn't come off as successfully as it had done on the larger locomotives. The sweep of the front-end casings was to similar proportions as that of the A4s and this gave the impression that the smokeboxes accounted for almost one-third of the locomotives' length. The wedge-front cabs and the rear sweep of the wheel valances could not disguise the fact that, at the rear of the locomotives, there were 6ft 8in driving wheels and not dainty pony trucks. The streamlining of the 4,200 gallon tenders added to the overall image of all front and rear with no real middle. During the war years, the pair of streamlined B17s followed the lead of the A4s and had their lower skirtings removed in order to simplify maintenance and, in 1951, the rest of the casings were removed completely.

Apart from the P2s and the pair of B17s, there was one other LNER locomotive which received the full A4-style streamlining and this was Gresley's old embarrassment, the W1 4-6-4 No 10000. The Galloping Sausage had had its own form of streamlining from the time it had been built in 1929 but,

The well-polished streamlined B17 No 2859 *East Anglian* was photographed near Ipswich on the London-bound East Anglian in the autumn of 1937.

Photo: Rail Archive Stephenson

When viewed from the front, the streamlining which was applied to the pair of B17 4-6-0s was very pleasing. The main purpose of streamlining the B17s was to publicise the East Anglian service to Norwich which was inaugurated on 27 September 1937 and, here, No 2859 *East Anglian* is being turned at Liverpool Street on 28 September, the second day of the new service.

Photo: LCGB/Ken Nunn Collection

The side view of the streamlined B17s made them look rather stunted but, nevertheless, the apple green livery was rather nice. This picture shows No 2870 *City of London* near Shenfield with the down East Anglian on 10 June 1938.

Photo: LCGB/Ken Nunn Collection

despite its revolutionary appearance, its ability to remain in service for any length of time had, quite frankly, underwhelmed. To the delight of crews and shedmasters alike, the locomotive had been taken out of service on 21 August 1935 for storage at Darlington and, much to the relief of the staff there, it was taken to Doncaster works on 13 October 1936.

Its journey from Darlington to Doncaster was made under its own steam and, in view of the loco-motive's employment record, no chances were taken for the epic trip which was all of seventy-six miles. Apart from the usual footplate crew, two fitters travelled with the locomotive and the party was issued with firm instructions that, in the event of a breakdown, no other personnel were to drive the locomotive nor was it to be left unattended. In order to allow for any possible eccentricities that No 10000 might display en route, no time limit was placed on the trip.

On its intrepid journey to Doncaster, No 10000 astounded the crew by its unwillingness to display any of its characteristic idiosyncrasies. Instead, it developed the completely new tendency for its bearings to overheat at regular intervals and, as a result, the trip took all of fifteen hours. The reason behind the resurrection of the locomotive and its trip to Doncaster works was that Gresley had decided to rebuild it as a three-cylinder non-compound locomotive with a boiler similar in size to those of the P2s albeit with 250lb pressure. The work was completed early in November 1937 and, apart from the complete mechanical transfor-mation, No 10000 was fitted with full streamlined casing in the A4 style and painted in Garter blue livery with red wheels.

A welcome consequence of the 250lb boiler and the use of three 20in diameter cylinders was that the tractive effort of the 'new' No 10000 was a macho 41,437lb. The rebuilding had incorporated the shortening of the front of the frames by eight-een inches, a new front bogie and the use of a Kylchap double blastpipe and chimney whereas the original corridor tender was retained with only minimal modifications. The rear frames remained unaltered and the removal of the water-tube boiler meant that the locomotive had a far more spacious cab than was the norm on the LNER. Despite its

completely new identity, the past reputation of No 10000 had not been forgotten and, when it was sent to Kings Cross shed on 6 November 1937, the depot staff faced its arrival with trepidation.

This was soon found to be unwarranted. It turned out to be a very useful acquisition for Kings Cross shed although, despite its hefty tractive effort, it was in no way superior to any of the 4-6-2 classes. Many of its early duties were between Kings Cross and Newcastle and it was quite often trusted with the southbound Flying Scotsman from Newcastle. There were, however, the odd niggles which included overheated axleboxes and so, as a precaution, in February 1938 it was transferred to Doncaster shed where workshop facilities were nearby. This was not, fortunately, to prove a repeat of its earlier history and so, in March 1939, it was returned to Kings Cross from where it went on to perform satisfactorily on a variety of turns including the occasional outings on the Silver Jubilee and the Coronation.

For two weeks early in the summer of 1942, No 10000 was allocated to Haymarket shed where it was evaluated over the old problem route to Aberdeen. By this time, Edward Thompson was in control of locomotive matters on the LNER but, had No 10000's move to Scotland occurred during Gresley's time, much interesting speculation would have resulted. Gresley's usual policy was to con-struct just a small number of any newly-designed locomotives and to assess their performances before proceeding with the building of any others and, when No 10000 was completely rebuilt in 1937, there were rumours that it might prove to be a forerunner of a whole class. The long wheelbases of the P2s had proved problematical on the route between Edinburgh and Aberdeen and so the shorter wheelbase and the high nominal tractive effort of the 4-6-4 made it, on paper, an ideal candidate for the Aberdeen run. Gresley had never got round to testing No 10000 northwards from Edinburgh but, had he done so, he might have considered it worth-while to have built similar locomotives either to augment or replace the P2s on those duties.

When Thompson succeeded Gresley, the demands of wartime traffic required locomotives which dis-played strength more than speed but, inexplicably, the idea of transferring the powerful P2s to more

Much to the relief of crews and shedmasters alike, the unique water-tube W1 4-6-4 No 10000 was put into storage in August 1935. The following year, it was sent to Doncaster works and rebuilt as a conventional three-cylinder locomotive but with the streamlined casing which had brought such attention to the A4s. This picture of the rebuilt W1 was taken at Doncaster works yard in 1939.   *Photo: Rail Archive Stephenson*

TABLE 5.2: ORIGINAL DIMENSIONS OF REBUILT W1 CLASS 4-6-4 No 10000

REBUILT:	Doncaster 1937
WEIGHTS FULL (locomotive):	107 tons 17 cwt
(tender):	64 tons 3 cwt (a)
TENDER CAPACITY (water):	5,000 gallons
(coal):	8 tons (a)
WHEELBASE (locomotive):	40ft 0in
(tender):	16ft 0in
WHEEL DIAMETERS (leading):	3ft 2in
(coupled):	6ft 8in
(trailing):	3ft 2in
CYLINDERS:	(3) 20in × 26in (b)
BOILER PRESSURE:	250 lb
HEATING SURFACES (tubes):	1281.4 sq ft
(flues):	1063.7 sq ft
(firebox):	252.5 sq ft
(superheater):	748.9 sq ft
SUPERHEATER:	Robinson 43-element
GRATE AREA:	50 sq ft
TRACTIVE EFFORT @ 85% boiler pressure:	41,437 lb (b)
LNER ROUTE AVAILABILITY:	9
BR POWER CLASSIFICATION:	7P (8P from Jan. 1951)

Notes:

(a) Tender modified in January 1938. From then, full weight was 64 tons 3 cwt and coal capacity was 9 tons.

(b) In December 1956, cylinders of 19in × 26in were fitted and this reduced the tractive effort to 37,397lb.

Pictures of the W1 4-6-4 working well were a rarity when it ran as a compound water-tube machine but its rebuilding as a non-compound engine transformed it into yet another reliable piece of Gresleyana. This photograph was taken on 10 May 1938 near Potters Bar and shows the engine exuding supreme confidence at the head of the 4.00pm Kings Cross to Newcastle express.

Photo: LCGB/Ken Nunn Collection

Like its celebrated cousins, the A4s, W1 4-6-4 No 10000 lost its lower skirting during the war years and it was never replaced. As BR No 60700, the locomotive is seen leaving Huntingdon on 14 August 1952 with a northbound express.

Photo: E. H. Sawford

compatible duties on the East Coast main line was not considered. Even if No 10000 had turned out to be an ideal replacement for the P2s on the Aberdeen route, it would have been highly unlikely that Thompson would have rushed out to order a batch of similar locomotives and, as he seemed not to have considered the potential of the P2s in England, it is difficult to see what he had hoped to achieve by trying No 10000 on the P2's traditional territory. On duties between Edinburgh and Aberdeen, No 10000 proved no better or worse than the P2s and this lack of conclusiveness was, perhaps, the ideal result for Edward Thompson.

Like all other LNER express locomotives, No 10000 did not escape the black livery of the war years and it was December 1946 before it was restored to the glory of Garter blue. As with the other streamlined LNER locomotives, the lower skirting was removed during the war in order to facilitate the routine maintenance of its running gear, and it was never replaced. By the end of the war, the locomotive was firmly ensconced back at

Kings Cross shed after its brief Scottish excursion and, from its London base, its main sphere of operation was the East Coast main line although it was known to make occasional appearances at Leeds and Cambridge. Despite the power of the locomotive, it was not required to repeat its wartime exploits of hauling trains in excess of 700 tons. Under British Railways ownership, it was re-numbered 60700 and, in January 1951, it received its new livery of ultramarine with black and white lining. It was originally intended that, after re-painting, it would be named *Pegasus* but, although the nameplates were cast, they were never applied. The locomotive's final change of livery was the standard British Railways dark green with orange and black lining and this was applied in May 1952. From 25 October 1953, No 60700 was allocated to Doncaster shed. From there, many of its duties were southwards to Kings Cross and it notched up some excellent performances at the head of express trains. In the late 1950s, however, its express turns became less and less and, instead, it was more

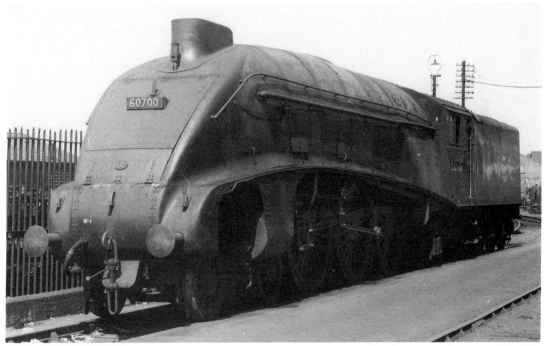

The unique W1 4-6-4 was photographed at Doncaster works yard in May 1959 and its 36A shedplate is clearly visible. The engine is showing signs of neglect and, considering that it was withdrawn on the first of the following month, the cleaners presumably knew about the engine's impending fate.

Photo: Rail Archive Stephenson

TABLE 5.3: ALLOCATIONS OF REBUILT W1 CLASS 4-6-4 No 10000

11/1937:	Kings Cross	3/1939:	Kings Cross	6/1942:	Kings Cross
2/1938:	Doncaster	5/1942:	Haymarket	10/1953:	Doncaster

Withdrawn: 1/6/1959

often used on stopping trains to Peterborough, Leeds or Sheffield. Occasionally, No 60700 was put to work on braked goods trains but, as it had been known to haul the odd coal train or two in the immediate post-war years, these inglorious turns were not completely foreign to it.

On 1 June 1959, No 60700 was withdrawn after an eventful life of almost thirty years. It had started out as Nigel Gresley's boldest experiment to date but had soon become his greatest embarrassment, since the Great Western's Castle class 4-6-0s had knocked spots off his A1 4-6-2s in the exchange trials of 1925. Since the earliest days of the railways, many locomotives and classes had been treated to nicknames by both the crews and amateur enthusiasts, and this had usually been done out of either respect or affection. In the case of the 4-6-4, however, there was little respect or affection involved in the name of the 'Galloping Sausage'.

Gresley had shown inordinate patience with his pet and had put up with a variety of comments before deciding to take it off the road in 1936. By that time, of course, his A4s had fully restored his reputation, and the experience gained from not only the A4s but also the second batch of P2s helped to bring about the transformation of the Galloping Sausage into a successful locomotive. In a perverse way, Edward Thompson was seen to endorse the usefulness of the rebuilt version as, despite his wish to dispense with as much of Gresley's legacy as possible, he left the 4-6-4 alone. From the point of view of footplate crews, a duty

on No 10000 before it was rebuilt usually resulted in the driver and fireman telling their families not to wait up for them. After the rebuilding, however, the locomotive became a great favourite and several drivers even considered it the best machine on the LNER fleet although, with its grate area of fifty square feet, it took a little longer for the fireman to come to love it quite as dearly.

It is easy to poke fun at locomotives which didn't make the grade but, to be fair, no truly great engineer had a career which was free from the odd failure. Even George Churchward, the Great Western's celebrated chief mechanical engineer from 1902 to 1921, was lumbered with 4-6-2 No 111 *The Great Bear* which, admittedly, was neither of his own choosing nor anywhere near as unsatisfactory as Gresley's No 10000 but, nevertheless, was more of a white elephant than an ursine giant to its operating department. On the LMS, Fowler's 4-6-0, No 6399 *Fury*, didn't even enter revenue-earning service and, furthermore, while Gresley's 4-6-4 was, unquestionably, a magnificent-looking machine, *Fury* looked as it if had been built out of an ironmonger's surplus stock. In later years, the Southern Railway had its own conspicuous failure in the form of the Leader class express tank locomotive. Despite his Gresley upbringing and his successful 4-6-2s, Oliver Bulleid's one and only completed Leader had to face the media which, much to his acute discomfort, was far more developed in 1949 than at the time of Gresley's No 10000 in the early 1930s.

It is not just certain locomotives which have acquired the status of legends. The former Midland Railway line between Settle and Carlisle is, undoubtedly, one of the best-loved routes anywhere in Britain today and, naturally, it quickly became a magnet for the operation of preserved steam engines. This picture shows the wonderful bleakness of Ais Gill in the extreme north-west corner of Yorkshire. By the way, the locomotive is V2 No 4771 *Green Arrow* and it is seen heading the southbound Cumbrian Mountain Express on 30 September 1989.

Photo: Peter Herring

The only representative of Gresley's excellent V2 2-6-2s to be saved for preservation was No 60800 *Green Arrow* which was restored to LNER livery and received its original number, 4771. Here, it is seen at Marylebone which, with great respect to the inhabitants of NW1, does not really have the same visual appeal as Ais Gill.

Photo: Peter Herring

The nameplate of Gresley's K4, No 3442, was preserved in full working order and, when it was purchased by the preservationists, the entire engine was thrown in as a bonus. Here, *The Great Marquess* presides over Bridgnorth in October 1991.

Photo: P. Chancellor

CHAPTER 6

Green Arrow, Green Howard and the Snapper

The leader of the second batch of Gresley's P2 2-8-2s, No 2003 *Lord President*, was delivered in June 1936 and, that same month, the first of the V2 2-6-2s was completed at Doncaster. The preliminary drawings for the V2s had been made in August 1934 but the roots of the new locomotives could be traced back to 1932.

The building programme for 1932 included the extensive modification of a pair of Gresley's K3 2-6-0s, the design of which had originated on the

Great Northern Railway in 1920. The proposals for revamping the K3s involved major reconstruction as, not only were the 5ft 8in driving wheels to be replaced by ones of 6ft 2in, but also an articulated bogie was to be used to join the cab and the tender. The idea behind this unusual arrangement was to give a smoother ride at higher speeds and, previously, experiments had been conducted on a pair of Vincent Raven 4-4-2s, C7 (later C9) class Nos 727 and 2171. The articulated configuration which had

The first of Gresley's V2 2-6-2s, No 4771 *Green Arrow*, was delivered in June 1936 and, on the eighth of the following month, it was photographed ascending Belle Isle bank with the 5.00pm Kings Cross to Peterborough train.

Photo: LCGB/Ken Nunn Collection

been applied to the 4-4-2s had been, primarily, to augment the use of boosters but, although there was no intention to use boosters on the modified K3s, the tests yielded information which was very useful for future reference.

One of the major problems with the original plans for rebuilding the K3s was that the larger driving wheels would raise the profile of the locomotive to such an extent that, in order to keep within the limits of the loading gauge, the chimney could be no more than six inches high. With such a short chimney, there was little chance that exhaust smoke could be kept clear of the driver's field of vision and, as it was considered somewhat safer if a driver could see such trivial things as signals and the like, a system of air ducts was devised to lift the smoke clear of the cab. Drifting smoke was a constant problem of most LNER designs, despite the fact that David Jones had cracked the problem on the Highland Railway some fifty years earlier by the use of louvred chimneys. When the design work for the modified K3s was transferred from Doncaster to Darlington, the new draughtsmen found a way of lowering the pitch of the boiler so that a full-length chimney could be incorporated without the need for additional ducting.

The exhaust problem was just one of several aspects of the K3s which were the subject of extensive design work but, in October 1933, permission to proceed to the construction stage was cancelled. This caused a great degree of disappointment in the classification department, as the wheel arrangement which had been proposed for the rebuilt K3s could have been taken to be either a 2-6-0 or a 2-6-4-4. The opportunity to introduce a sub-class, or even a completely new class, had been lost.

Nevertheless, a new type of locomotive was still required for fast mixed traffic duties. This was no reflection on the K3s as they performed excellent work on secondary duties and, as if to endorse their usefulness, their construction was to continue until 1937. Under heavy loadings however, they were not the speediest of machines. Having been denied the chance to play with the pair of K3s, Gresley's answer to the gap in the motive power range was the V2 2-6-2s, but their final design showed major differences to the initial drawings of August 1934. Originally, Gresley had based the

design for V2s on that of P2 2-8-2 No 2001 *Cock O' The North* which had emerged in May 1934 and the proposed specifications of the new 2-6-2 locomotives included Lentz rotary gear, an A.C.F.I. feed-water heater, a Kylchap double exhaust system and a semi-streamlined front-end with integral smoke deflectors. A later amendment went the whole hog with the streamlining and incorporated the full casing which was, at the time, being developed for the A4s.

The final drawings for the V2s were somewhat different to the earlier ones. They dispensed with, not only all the streamlined casing, but also the Lentz gear and feed-water heater and so, when the locomotives eventually appeared, they gave the appearance not of modified P2s or A4s but, instead of slightly scaled-down A3s. This design retained the 6ft 2in driving wheels which had been included in the plans for the 'improved K3' and also the provision for casting the three 18½in × 26in cylinders, the steam chests and the saddles in one block. It was considered unnecessary to design new tenders for the V2s as the LNER's standard issue 4,200 gallon models had proved versatile enough to be used behind almost everything from J38 0-6-0s to O2 2-8-0s.

Gresley displayed his characteristic preference for an initial order of just a small number of locomotives of a new design and, after the first V2 was completed in June 1936, only four more were to emerge from Doncaster works by November. The standard LNER practice was to allocate random blank numbers to new locomotives and the one which was earmarked for the new V2 was No 637. The number was applied to the locomotive but, even before the engine embarked on trials, it was renumbered 4771. The change of heart over the locomotive's number came about when the LNER realised that it had a potentially numerous class on its hands and the new number had a lengthy run of consecutive blanks following on its heels. The four other V2s had had Nos 639/40/64/70 reserved for them but, following the lead of No 4771, they took Nos 4772 – 75 instead.

As No 4771 was destined for mixed traffic duties, its livery should, theoretically, have been black. The LNER, however, always had a corporate eye for a new marketing angle and this resulted in the

The fourth V2 to be delivered was No 4774. Here, it is seen near Potters Bar on 18 September 1937 with the 4.05pm Kings Cross to Cleethorpes train. This locomotive eventually became BR No 60803 and survived until June 1963.

Photo: LCGB/Ken Nunn Collection

TABLE 6.1: ORIGINAL DIMENSIONS OF V2 CLASS 2-6-2s

BUILT:	Doncaster and Darlington 1936–1944
WEIGHTS FULL (locomotive):	93 tons 2 cwt
(tender):	51 tons 0 cwt/52 tons 0 cwt (a)
TENDER CAPACITY (water):	4,200 gallons
(coal):	7 tons 10 cwt
WHEELBASE (locomotive):	33ft 8in
(tender):	13ft 6in
WHEEL DIAMETERS (leading):	3ft 2in
(coupled):	6ft 2in
(trailing):	3ft 8in
CYLINDERS:	(3) 18½ × 26in
BOILER PRESSURE:	220 lb
HEATING SURFACES (tubes):	1211.57 sq ft
(flues):	1004.50 sq ft
(firebox):	215 sq ft
(superheater):	679.67 sq ft
GRATE AREA:	41.25 sq ft
TRACTIVE EFFORT @ 85% boiler pressure:	33,730 lb
LNER ROUTE AVAILABILITY:	9
BR POWER CLASSIFICATION:	6MT

Note:

(a) Heavier tenders introduced in 1941.

locomotive being turned out in smartly-lined apple green. The reason for the upmarket livery was that, at the time, the company's express parcels service was marketed under the banner of 'Green Arrow' and this prompted not only the green livery but also the naming of the locomotive. The original nameplates of *Green Arrow* were in the curved style as used on the 4-6-2s and were mounted on No 4771's running plates, but these nameplates were soon replaced by straight ones which were mounted on the sides of the smokebox. The naming of No 4771 as *Green Arrow* did not establish a precedent, as the LNER did not normally name freight or mixed traffic locomotives and, consequently, the other five V2s which were built in 1936 retained their anonymity.

Of the first five V2s, No 4771 was allocated to Kings Cross while York received Nos 4772/73; New England shed at Peterborough took No 4774 and the remaining one, No 4775, was despatched to Dundee. Initially, Kings Cross shed kept No 4771 mainly on express freight turns as far as Peterborough although, during the peak summer period, the intensity of Saturday passenger traffic and excursion work required its attentions. Many of the freight duties which were worked by the New England V2, No 4774, had originated at Kings Cross and had been hauled to Peterborough by its classmate, No 4771. From its New England base, No 4774 worked both northwards to York and southwards to Kings Cross and, in common with its colleague in London, took its share of peak-season passenger turns. The sole Scottish V2, No 4775, worked northwards to Aberdeen and southwards to Edinburgh but, unlike its English pals, its activities on passenger duties included regular diagrammed work and not just Saturday and excursion jaunts.

The thinking behind the wide dispersal of the new locomotives was that, as they were intended for mixed traffic duties, they could be assessed in a variety of situations. The V2s might not have had

the same striking looks of the A4s or the impressive proportions of the P2s but, in professional circles, it was suggested that the V2s seemed to be the best locomotives that Gresley had designed. They benefited from the wealth of experience which Gresley had accumulated in his eventful career and, consequently, they incorporated the features which had been found to be most useful while dispensing with unnecessary frippery. Had the Kylchap double exhaust system been in a more advanced state of development, it would no doubt have been an automatic addition. The only two real drawbacks of the design of the V2s were, firstly, the axle weight of twenty-two tons and, secondly, that the wheelbase was only twenty-five inches less than that of the A3s. These factors prevented their use over a total of 57% of the LNER's network and these no-go areas included a number of important subsidiary lines and, unfortunately, most of the former Great Eastern section.

The activities of the first five V2s were scrutinised very closely and their performances did not disappoint. As a result, Gresley had no hesitation in requesting the construction of further V2s and two orders for a total of sixty-seven new locomotives were placed even before Nos 4771–75 had been fully assessed in the field. These new V2s were all built at Darlington works and were completed between July 1937 and September 1939. Further orders for a total of eighty-one V2s were placed between July 1938 and January 1940; Darlington turned out sixty-one of these with Doncaster producing the other twenty and all were delivered between April 1939 and July 1942. The eagle-eyed enthusiast or traditionalist could easily differentiate between the Doncaster and Darlington V2s as the former followed the old Great Northern penchant for having black cylinder covers whereas the latter followed the North Eastern tradition of having green covers.

The success of the V2s was reflected not only by

TABLE 6.2: CONSTRUCTION OF V2 2-6-2s

All locomotives of the class were constructed at Darlington except the following which were built at Doncaster:
LNER Nos 4771–4775 (BR Nos 60800–60804)
LNER Nos 4843–4852 (BR Nos 60872–60881)
LNER Nos 3655–3664 (BR Nos 60928–60937)

From new, V2 No 4790 was allocated to Ferryhill shed at Aberdeen and thereby claimed the distinction of being only the third of the class to find a home north of the border. This picture was taken at Aberdeen and is officially dated 1937 but, as No 4790 was not delivered until November that year, the high overhead sun must have been a winter phenomenon.

Photo: D. K. Jones Collection

their abundance but also by the attitude of Gresley's successor, Edward Thompson. Try as he might to find fault with the Gresley machines, Thompson had little alternative but to acknowledge the fact that the V2s were excellent locomotives and, although he must have found it painful, he ordered a batch of thirty-five more. Of these, only thirty-one were actually constructed as the order for the last four was amended to enable the building of a quartet of A2/1 class 4-6-2s. The V2s which Thompson ordered were all built at Darlington; the first was completed in August 1942 and the last in July 1944, just over eight years after No 4771 had made its debut. The delivery of the last V2 brought the total number of locomotives in the class to 184 and their numbers ran from 4771 to 4899 and 3641 to 3695.

As the V2s became more abundant, they took over more and more of the turns previously handled by the K3 2-6-0s and the C2 class Ivatt 4-4-2s and, although the K3s were found plenty of alternative

work on secondary duties, there was little scope for the ageing C2s and so they were gradually retired. The five original V2s had proved that they felt at home in any area and so, as subsequent members of the class appeared, the new arrivals were scattered throughout much of the LNER's network from Kings Cross to Aberdeen. Until the end of 1939, the former Great Northern section received over half of all new V2s but, between 1940 and 1944, the greatest number went to the former North Eastern section. By the time all 184 V2s were delivered, the ex-Great Northern section had 88, the former North Eastern section had 71 and the remaining 25 were stationed in Scotland.

The V2s excelled on fast freight duties and loadings in excess of 600 tons were far from unknown. When they were called upon for passenger turns, they proved to be highly versatile as they could handle not only relief and excursion loadings with ease, but also express turns when the need arose. In the event of a failure of a 4-6-2, shedmasters had

TABLE 6.3: THE V2 CLASS 2-6-2s. (Locomotives built 1936 – 38.)

FIRST NO.	BR NO.	BUILT	ALLOCATION 1/4/58	WDN.	FIRST NO.	BR NO.	BUILT	ALLOCATION 1/4/58	WDN.
4771	60800	6/36	Kings Cross	8/62	4793	60822	12/37	Dundee	12/64
4772	60801	8/36	Tweedmouth	10/62	4794	60823	12/37	St. Margarets	3/62
4773	60802	10/36	Heaton	3/64	4795	60824	12/37	Ferryhill	9/66
4774	60803	10/36	March	6/63	4796	60825	1/38	St. Margarets	4/64
4775	60804	11/36	Dundee	12/63	4797	60826	1/38	New England	4/62
4776	60805	7/37	Gateshead	12/63	4798	60827	1/38	Ferryhill	12/62
4777	60806	8/37	Heaton	9/66	4799	60828	3/38	Leicester	10/65
4778	60807	8/37	Gateshead	11/62	4800	60829	4/38	New England	5/62
4779	60808	8/37	Tweedmouth	10/64	4801	60830	4/38	March	6/63
4780	60809	8/37	Gateshead	7/64	4802	60831	5/38	Leicester	12/66
4781	60810	9/37	Heaton	11/65	4803	60832	7/38	New England	12/62
4782	60811	9/37	Heaton	4/62	4804	60833	8/38	Gateshead	5/64
4783	60812	9/37	Heaton	7/64	4805	60834	9/38	Dundee	3/64
4784	60813	9/37	St. Margarets	9/66	4806	60835	9/38	Heaton	10/65
4785	60814	10/37	Kings Cross	4/63	4807	60836	9/38	St. Margarets	12/66
4786	60815	10/37	Woodford Hs	4/62	4808	60837	10/38	York	11/65
4787	60816	10/37	St. Margarets	10/65	4809	60838	10/38	Dundee	1/64
4788	60817	11/37	Doncaster	6/63	4810	60839	10/38	York	10/62
4789	60818	11/37	St. Margarets	8/66	4811	60840	10/38	St. Margarets	12/62
4790	60819	11/37	Ferryhill	12/62	4812	60841	11/38	Doncaster	9/63
4791	60820	11/37	Kings Cross	6/62	4813	60842	12/38	Leicester	10/62
4792	60821	12/37	Grantham	12/62	4814	60843	12/38	York	10/65

no hesitation in using the V2s as replacements even on such high-profile trains as the Silver Jubilee, the Coronation, the West Riding Limited and the Yorkshire Pullman. One recorded replacement duty involved No 4789 which, in 1939, took the streamlined West Riding non-stop over the 185 ¾ miles from Kings Cross to Leeds in 168 minutes, just four minutes outside the schedule for the A4s. Another notable exploit featured No 4817 which notched up 93 mph while hauling the Yorkshire Pullman. On the former North Eastern section, the V2s became the largest locomotives to be used between York and Scarborough as, although the 4-6-2s were not officially barred from the route, the turntable at Scarborough could not accommodate them and it was rather too long a jaunt back to York to consider returning tender-first.

Of the V2s which were allocated to Scotland, St. Margaret's shed in Edinburgh became home to more than half of them. Most of the others tended

to live at Aberdeen or Dundee although Haymarket (Edinburgh), Eastfield (Glasgow) and Carlisle also received new V2s. In Scotland, the pattern set by No 4775 in 1936 was maintained and the V2s were regarded as true mixed traffic locomotives rather than as goods locomotives which could, if necessary, be forced on to other duties. Their regular work included express passenger turns between Aberdeen and Edinburgh and from Edinburgh to Glasgow and Berwick, as well as the express fish trains southwards from Aberdeen. When working from Edinburgh to Aberdeen, the V2s were allowed the same 480-ton loadings as the A1 and A3 4-6-2s while, in the return direction, the 420-ton limit from Aberdeen to Dundee and 450-ton restriction from Dundee to Edinburgh were both, once again, identical to those of the 4-6-2s. The V2s were also welcomed on the tortuous Waverley route between Edinburgh and Carlisle where their actual loadings on passenger trains were usually the same as those

Table 219.

EDINBURGH, CARLISLE AND LONDON (St. Pancras).

A full-page railway timetable for the route Edinburgh, Carlisle and London (St. Pancras), with columns for WEEKDAYS and SUNDAYS services.

Stations (reading down):

- Inverness (via Aberdeen) — dep.
- Inverness (via Dunkeld)
- Perth
- Stirling
- Aberdeen
- Dundee (Tay Bridge)
- Glasgow (Queen Street)
- EDINBURGH (Waverley) — dep.
- Abbeyhill
- Piershill
- Portobello
- Niddrie
- Eskbank
- Newtongrange
- Gorebridge
- Fushiebridge
- Tynehead
- Heriot
- Fountainhall
- Stow
- Bowland
- Galashiels
- Melrose
- St. Boswells
- Kelso
- St. Boswells — dep.
- Belses for Ancrum and Lilliesleaf
- Hassendean
- Hawick — arr.
- Hawick — dep.
- Stobs
- Shankend
- Riccarton Junction — arr.
- Riccarton Junction — dep.
- Newcastle
- Steele Road
- Kershope Foot
- Penton
- Riddings
- Scotch Dyke
- Longtown
- CARLISLE (Citadel) — arr.
- Carlisle (Citadel) — dep.
- Leeds (City)
- Bradford (Forster Square)
- Sheffield
- Nottingham
- Leicester
- Birmingham (New Street)
- Bristol (Temple Meads)
- LONDON (St. Pancras) — arr.

Footnotes:

A For other trains between Edinburgh and Eskbank see Table 236.
B On Sundays arrives Leeds 3.6 and Sheffield 4.11 a.m.
C On Saturdays arrives Sheffield 3.38 a.m.
D On Sundays arrives Birmingham 8.20 a.m. and Bristol 12.5 p.m.
E On Sundays arrives St. Pancras 6.42 p.m.
F Saturdays and Sundays excepted.
G On Saturdays runs 3 minutes later.
H On Saturdays arrives Newcastle 12.19 p.m.

J Commencing 1st June leaves Stirling 8.25 p.m.
K Leaves Carlisle 12.40 p.m.
N Calls at Hassendean when required to set down passengers on notice being given to the guard.
R Not after 28th May.
S Commences 5th June.
SO Saturdays only.

SX Saturdays excepted.
WSO Wednesdays and Saturdays only.
X One class only.
Y One class only except on Saturdays.
a a.m.
c On Saturday 2nd July leaves Dundee 6.20 p.m.
p p.m.

The Waverley Route between Edinburgh and Carlisle was one of the eternal problem lines for the operating department. This LNER timetable is for the summer of 1938 and, at that time, D49 4-4-0s, A3 4-6-2s and V2 2-6-2s were among the locomotives to be put to the test.

Throughout their useful lives, the V2s were subjected to very few changes, except for the obligatory black livery which was applied during the war. This picture shows No 60874, complete with 35A (New England) shedplate on 16 July 1951 and its black livery retains a matt finish courtesy of lackadaisical polishing.

Photo: E. R. Morten

given to the 4-6-2s.

Although none of the early V2s was allocated from new to sheds on the former Great Central network, they started to make occasional appearances at Marylebone as early as 1937. Their axle weights precluded their use on some sections of the ex-Great Central system but the main line through Leicester and Rugby presented no problem although, until seventy-foot turntables were installed at Marylebone and Leicester in 1938, the use of a V2 on the ex-Great Central main line was usually considered more trouble than it was worth. The first V2 to be formally transferred to an ex-Great Central shed was No 4798 which was allocated to Gorton in October 1938, primarily to share duties on express passenger services to Marylebone.

In 1939, three more V2s were transferred to Gorton and their appearances on freight duties were heavily outnumbered by those on passenger workings. The former Cheshire Lines Committee route from Manchester (Central) to Liverpool

(Central) had always been considered out of bounds for the V2s because of their axle weights but, in July 1939, the civil engineers had a rethink and No 4798 was put in charge of a scheduled train over the route. Seemingly oblivious to the normally tentative handling of test workings, and ignoring the white knuckles of the civil engineers, No 4798 was recorded at a speed of 79 mph between Irlam and Padgate. The locomotive had to be turned at Brunswick shed before its return trip and although the turntable was, in theory, of adequate length, a section had to be hacked out of a stone wall to enable the operation to be carried out. Unfortunately, the V2s reign on Manchester to Liverpool expresses was brought to an early end by the outbreak of war, as the Gorton-based V2s were transferred for heavy duties elsewhere.

The class leader, No 4771 *Green Arrow*, was the only V2 to be named from new but, between September 1937 and June 1939, six others received names at special ceremonies. On 11 September

1937, No 4780 was given the name *The Snapper* which was appended with *The East Yorkshire Regiment, The Duke of York's Own*. The naming ceremony was conducted by the Colonel of the Regiment, Brigadier General J. L. J. Clarke, at Hull Paragon station and the newly-named locomotive hauled the train of dignitaries from Hull back to London. Seemingly unimpressed with its new handle, No 4780 broke down at Beverley.

As if the casting of the nameplates for No 4780 hadn't provided enough work for the workshop lads, on 24 September 1938 No 4806 was named *The Green Howard, Alexandra, Princess of Wales's Own Yorkshire Regiment*. The naming ceremony, which was conducted at Richmond station, marked the 250th anniversary of the founding of the regiment. By this time, the staff in the LNER's nameplate-casting department had come to associate the ceremonial naming of a V2 with a welcome bit of overtime but, much to their chagrin, the next naming of a V2 turned out to require less of the alphabet. During a ceremony at York station on 3

April 1939, No 4818 was named, quite simply, *St Peter's School*.

The next V2 to be named was No 4843 which was christened on 20 May 1939. The name which was applied to the locomotive was *King's Own Yorkshire Light Infantry* and, while it was a lengthy name when compared to those of most other locomotives, it was, by the standards of the V2s, remarkably brief. As if to reflect that the naming of a V2 was nothing new, the LNER chose to conduct the ceremony in the yard of Doncaster works. The Colonel of the Regiment didn't seem too impressed either, as he sent along his wife to officiate on his behalf. It was back to the short nameplates for the next V2 to be rescued from anonymity. On 15 June 1939, No 4831 was named *Durham School* in a ceremony at Elvet station in Durham and, had any of the pupils or staff of the school anticipated combining the formalities with a bit of train-spotting, they would have been rather disappointed. Elvet had been closed to passengers from 1 January 1931 and, since then, had been used solely for

What a difference a clean makes. V2 No 60937 stands outside Darlington works on 7 July 1956 fresh from an overhaul; it is soon to return to its home shed at Dundee.

Photo: E. H. Sawford

TABLE 6.4: THE V2 CLASS 2-6-2s. (Locomotives built 1939 – 40.)

FIRST NO.	BR NO.	BUILT	ALLOCATION 1/4/58	WDN.	FIRST NO.	BR NO.	BUILT	ALLOCATION 1/4/58	WDN.
4815	60844	2/39	Dundee	11/65	4852	60881	10/40	Doncaster	6/63
4816	60845	2/39	New England	9/62	4853	60882	10/39	St. Margarets	7/64
4817	60846	2/39	Ardsley	10/65	4854	60883	10/39	St. Margarets	2/63
4818	60847	3/39	York	6/65	4855	60884	10/39	Ardsley	9/65
4819	60848	3/39	York	7/62	4856	60885	11/39	Copley Hill	9/65
4820	60849	3/39	Doncaster	4/62	4857	60886	11/39	Heaton	4/66
4821	60850	3/39	New England	2/62	4858	60887	11/39	York	7/64
4822	60851	3/39	Ferryhill	12/62	4859	60888	12/39	Ferryhill	12/62
4823	60852	3/39	Doncaster	9/63	4860	60889	12/39	Doncaster	6/63
4824	60853	4/39	New England	9/63	4861	60890	12/39	Woodford Hs	4/62
4825	60854	4/39	Kings Cross	6/63	4862	60891	12/39	Heaton	10/64
4826	60855	4/39	Neasden	4/64	4863	60892	1/40	St. Margarets	11/63
4827	60856	5/39	York	5/64	4864	60893	1/40	Grantham	9/62
4828	60857	5/39	Doncaster	4/62	4865	60894	1/40	St. Margarets	12/62
4829	60858	5/39	March	10/63	4866	60895	1/40	York	10/65
4830	60859	5/39	Copley Hill	9/65	4867	60896	2/40	Doncaster	9/62
4831	60860	5/39	Gateshead	10/62	4868	60897	2/40	New England	6/63
4832	60861	6/39	Ardsley	8/63	4869	60898	2/40	Ferryhill	11/63
4833	60862	6/39	Kings Cross	6/63	4870	60899	2/40	Doncaster	9/63
4834	60863	6/39	Leicester	4/62	4871	60900	3/40	St. Margarets	4/63
4835	60864	6/39	York	3/64	4872	60901	3/40	Heaton	6/65
4836	60865	6/39	Copley Hill	6/65	4873	60902	3/40	Kings Cross	9/63
4837	60866	7/39	New England	12/62	4874	60903	3/40	Kings Cross	2/63
4838	60867	7/39	New England	5/62	4875	60904	4/40	York	7/64
4839	60868	7/39	Gateshead	9/66	4876	60905	4/40	Doncaster	9/63
4840	60869	8/39	New England	6/63	4877	60906	4/40	New England	5/63
4841	60870	8/39	Doncaster	7/63	4878	60907	4/40	York	5/62
4842	60871	9/39	Kings Cross	9/63	4879	60908	4/40	New England	6/62
4843	60872	4/39	Doncaster	9/63	4880	60909	5/40	Doncaster	6/62
4844	60873	5/39	St. Margarets	12/62	4881	60910	5/40	Heaton	4/64
4845	60874	6/39	New England	8/62	4882	60911	5/40	Leicester	12/62
4846	60875	7/39	New England	3/62	4883	60912	5/40	New England	4/63
4847	60876	5/40	Neasden	10/65	4884	60913	6/40	Copley Hill	10/64
4848	60877	6/40	Neasden	2/66	4885	60914	6/40	Kings Cross	9/62
4849	60878	7/40	Leicester	10/62	4886	60915	6/40	Woodford Hs	11/62
4850	60879	8/40	Leicester	12/62	4887	60916	7/40	Ardsley	6/64
4851	60880	9/40	Doncaster	9/63	4888	60917	8/40	Doncaster	4/62

goods. The only V2 to receive a single-word name-plate was No 4844 which was named *Coldstreamer* on 20 June 1939 at a ceremony at Kings Cross station. There were plans to give the name *The Royal Grammar School, Newcastle Upon Tyne,*

A.D. 1545 to No 4804 in the autumn of 1939 but, because of the outbreak of war, the idea was shelved and never the saw the light of day again.

During the war, the V2s inevitably had their impressive green liveries replaced by unlined black.

Their versatility was utilised to the full and they played their part on everything from the Flying Scotsman to slow goods trains while, in the haulage stakes, they did not grumble about twenty-coach trains of over 600 tons. One of the class, No 4800, took an 850-ton passenger train from Peterborough to Kings Cross and this is believed to be the heaviest troop train ever to run on the LNER. That train comprised twenty-six bogie coaches and, when the locomotive drew to a halt at Kings Cross station, its last four coaches were still in Gasworks Tunnel. The South Yorkshire coal traffic placed heavy demands on local locomotive depots and, in 1940, Sheffield (Darnall) and Mexborough sheds both acquired their first V2s to augment their more traditional studs of heavy freight designs. The esteem in which the V2s were held was reflected by the fact that, apart from heavy freight engines, they were the only locomotives to be built in any quantity during the war.

After the war, it was decided that the class would retain its black livery although, in September 1946, one managed to defy authority and obtain apple green paintwork. It was not a long-lived transformation as, by July 1948, it was in black once more. The locomotive concerned was originally No 4854 which, at the time it received its green livery, had become No 883. Edward Thompson's renumbering scheme of 1943 allocated Nos 700–883 to the V2s but only nineteen managed to be painted with their new numbers before the 1946 renumbering scheme came up with Nos 800–983 for the class.

The year of 1946 was not a good one for the V2s. Seven of the class were involved in accidents and, although the mishaps were not disastrous, suggestions were raised that they were not just coincidences. It was revealed that, at the scenes of most of the accidents, the tracks were in poor condition even by the standards of the early post-war years but, despite this, other types of locomotives had not experienced the same difficulties.

Not many of the V2s were restored to green liveries after the compulsory wartime coating of black. One of the exceptions was Kings Cross-based No 60828 which is seen on 12 June 1957 pulling out of Huntingdon with a northbound express.
Photo: E. H. Sawford

123

TABLE 6.5: THE V2 CLASS 2-6-2s. (Locomotives built 1941/42.)

FIRST NO.	BR NO.	BUILT	ALLOCATION 1/4/58	WDN.	FIRST NO.	BR NO.	BUILT	ALLOCATION 1/4/58	WDN.
4889	60918	8/41	York	10/62	3642	60940	2/42	Gateshead	10/65
4890	60919	9/41	Ferryhill	9/66	3643	60941	3/42	York	7/64
4891	60920	10/41	Dundee	12/62	3644	60942	3/42	Gateshead	5/64
4892	60921	10/41	Doncaster	6/63	3645	60943	3/42	Doncaster	9/62
4893	60922	10/41	Heaton	7/64	3646	60944	4/42	Heaton	9/65
4894	60923	11/41	Gateshead	10/65	3647	60945	5/42	Heaton	7/64
4895	60924	11/41	New England	9/63	3648	60946	5/42	York	10/65
4896	60925	12/41	York	5/64	3649	60947	6/42	Gateshead	10/62
4897	60926	12/41	Tweedmouth	10/62	3650	60948	6/42	March	9/63
4898	60927	12/41	St. Margarets	12/62	3651	60949	6/42	Gateshead	11/62
					3652	60950	6/42	Kings Cross	9/63
3655	60928	6/41	Doncaster	3/62	3653	60951	7/42	St. Margarets	12/62
3656	60929	6/41	York	6/65	3654	60952	7/42	York	10/65
3657	60930	7/41	Doncaster	9/62					
3658	60931	8/41	Dundee	9/65	3665	60953	8/42	St. Margarets	5/62
3659	60932	10/41	Tweedmouth	5/64	3666	60954	9/42	York	11/63
3660	60933	10/41	St. Margarets	12/62	3667	60955	9/42	Ferryhill	9/66
3661	60934	1/42	York	10/62	3668	60956	9/42	Doncaster	9/62
3662	60935	2/42	Doncaster	6/63	3669	60957	10/42	Haymarket	12/64
3663	60936	3/42	New England	9/62	3670	60958	10/42	Dundee	12/62
3664	60937	3/42	Dundee	12/62	3671	60959	11/42	St. Margarets	7/63
					3672	60960	11/42	York	2/62
4899	60938	1/42	March	10/62	3673	60961	12/42	York	4/65
3641	60939	2/42	York	10/64	3674	60962	12/42	Heaton	9/65

Edward Thompson had no hesitation in pointing the finger at the Gresley-designed pony trucks of the V2s, and voiced his long-held opinion that the swing-link trucks were vulnerable on anything less than perfect track. Being a truly magnanimous character, Thompson quickly pointed out that the side-sprung pony trucks he had fitted to his L1 class 2-6-4Ts did not suffer from the same problems and so, in November 1946, V2 No 884 (ex-LNER No 4855) was similarly equipped. The rest of the class eventually followed suit.

Under the ownership of British Railways, the V2s were given Nos 60800–60983. At the time of Nationalisation, eighty-two of the class were divided among the three main ex-Great Northern sheds of Kings Cross, Doncaster and New England; a further fifty-three were shared by the ex-North Eastern depots of York and Gateshead while the Scottish contingent of thirty-three was split between Haymarket, Dundee and Aberdeen.

By the early 1950s, the operation of Britain's railways had returned to a reasonable degree of normality, albeit under state ownership, and the V2s had largely reverted to their old familiar duties. The locomotives' forays into new areas were, on the whole, subtle although the manifestation of No 60845 at Swindon between May 1952 and April 1953 was a little bit out of the ordinary. The Swindon visit was to enable the draughting of the locomotive's self-cleaning smokeboxes to be evaluated on the test plant at the works and, in order to test the machine thoroughly, trials were also conducted on the open road. The test ground was the line between Reading and Stoke Gifford, near Bristol, and the loadings included a twenty-five coach train of over 750 tons.

The results of the tests indicated that modifications to the blastpipe and chimney liner would provide the solution and Swindon carried out the work. The improvement in the locomotive's performance was immediately apparent and, subsequently, twenty V2s were similarly treated at Darlington. Self-cleaning smokeboxes had first appeared on LNER locomotives in 1946 and they followed the style of those used by the LMS. It may have been of some consolation to the folks at Doncaster and Darlington to know that, in common with the LNER's V2s, most of the LMS three-cylinder locomotives did not take well to the self-cleaning equipment.

The protracted stay of No 60845 in the West Country was a highly conspicuous departure from its customary haunts and, for those who were familiar with the workings of the Eastern Region, the sight of No 60835 *The Green Howard* on a special working at Ipswich in April 1953 was no less unusual. Apart from the V2s transferred to March shed in 1951 and 1952 for duties to Peterborough and beyond, No 60835's trip to Ipswich was one of only three known V2 appearances on East Anglian lines before the preservation era.

One month after *The Green Howard* had boldly gone where no V2 had gone before, six other members of the class went walkabout but, this time, well beyond the former LNER empire. All six were transferred to the Southern Region's shed at

Nine Elms in order to ease a sudden and severe shortage of motive power. After one of Bulleid's Merchant Navy class 4-6-2s suffered a fracture of its driving axle, examinations were made of all the Southern's other 4-6-2s and, alarmingly, a number of potentially dangerous axles were discovered. The offending locomotives were taken out of service immediately and sent for repair and, as this left the Southern Region with a dearth of express passenger locomotives, the three other Regions demonstrated the comradeship of state ownership and lent some of their own locomotives to the Southern. The Western and the Midland Regions provided Britannia class 4-6-2s while the Eastern sent V2s Nos 60893/96, 60908/16/17/28. The V2s spent about six weeks on the Southern and they worked out of Waterloo mainly to Bournemouth but, occasionally to Weymouth or Exeter. Generally, the Southern crews had difficulty in mastering the V2s and, consequently, late running of a V2-hauled train was not uncommon but, conversely, it was known for them to be pushed to 90 mph when in charge of the Bournemouth Belle.

During the latter half of the 1950s and the early 1960s, a number of V2s were treated to mechanical modifications which, apart from the replacement pony trucks and the Swindon-inspired smokebox treatment, were the only significant alterations made during the class's entire life. The first of the two changes involved replacing the original single-

TABLE 6.6: THE V2 CLASS 2-6-2s. (Locomotives built 1943/44.)

FIRST NO.	BR NO.	BUILT	ALLOCATION 1/4/58	WDN.	FIRST NO.	BR NO.	BUILT	ALLOCATION 1/4/58	WDN.
3675	60963	1/43	York	6/65	3686	60974	8/43	York	12/63
3676	60964	1/43	Gateshead	5/64	3687	60975	9/43	York	5/64
3677	60965	2/43	St. Margarets	12/62	3688	60976	10/43	York	9/66
3678	60966	3/43	Grantham	6/63	3689	60977	10/43	York	2/62
3679	60967	3/43	Gateshead	2/64	3690	60978	12/43	Heaton	11/62
3680	60968	4/43	York	5/63	3691	60979	1/44	Heaton	10/62
3681	60969	4/43	Dundee	5/64	3692	60980	3/44	St. Margarets	12/62
3682	60970	5/43	Ferryhill	2/66	3693	60981	4/44	York	4/63
3683	60971	6/43	Dundee	12/62	3694	60982	6/44	York	10/64
3684	60972	7/43	Ferryhill	11/63	3695	60983	7/44	Kings Cross	9/62
3685	60973	7/43	Ferryhill	1/66					

unit cylinders with three separate units, complete with external steam-pipes to the outside pair. This was done for reasons of economy, as it was felt that the practice of replacing an entire casting for all three cylinders and the steam chests was rather extravagant on the occasions when just one cylinder needed replacement and, between May 1956 and March 1962, a total of seventy-one V2s were modified.

The second mechanical change to be administered to the V2s around that time was the fitting of double blastpipes and chimneys in 1960 in an attempt to improve their steaming when using poor grade coal. Despite the undisputed efficiency of the Kylchap system and the previous experiences with appendages of LMS origin, it was decided to use LMS-style equipment on a pair of V2s but, as this was found to have no significant effect on the locomotives' steaming, the Kylchap system was fitted to a third V2 for comparison. This locomotive showed an immediate improvement in performance and so the order went out to convert the pair equipped with the LMS gear and also to fit the rest of the class with the Kylchap equipment. By this time, however, it was rather late in the day to think about significant alterations to an entire class of steam locomotives and only eight V2s actually received their double chimneys.

By the time the cylinder and exhaust modifications had been carried out, the V2s had all been treated to fully-lined Brunswick green liveries and this reflected the status of the class. British Railways had introduced the green livery in 1951 specifically for express passenger locomotives and, in 1956, the V2s were considered worthy of the same paintwork. On 29 April 1958, No 60964 received a name to go with its new colour scheme when, during a ceremony at Durham station, it was named *The Durham Light Infantry*.

The V2's excursions to Swindon and Nine Elms in the early 1950s were for specific purposes and, despite their weight of numbers and the lessening of the enforcement of boundary restrictions in the late 1950s and early 1960s, the sighting of a V2 outside of its home region in England was not particularly common. The official transfer of former Great Central lines south of Chesterfield to the Midland Region in 1958 resulted in the thirteen V2s which were allocated to Neasden, Woodford Halse and Leicester going on to the Midland's books, but this was hardly a case of the V2s exploring new territory. Members of the class were known to have worked through from York to Bristol on a few occasions and, in the early 1960s, they made sporadic forays to Derby and Birmingham, while the odd manifestation at St. Pancras, Clapham Junction, Barrow and Carnforth provided unforeseen attractions for local enthusiasts.

In Scotland, the old demarcation lines were less rigidly observed and V2s were used on scheduled passenger and freight workings over the former Glasgow & South Western's lines in 1960 and also on ex-Caledonian lines in 1963. The latter duties tended to be as substitutes for failed diesels, in particular the North British Type 2s which had first appeared in 1959. Despite the introduction of

TABLE 6.7: NAMING OF THE V2 2-6-2s.

LNER NO.	BR NO.	DATE NAMED	NAME
4771	60800	NEW	*Green Arrow*
4780	60809	11/9/1937	*The Snapper, The East Yorkshire Regiment, The Duke of York's Own*
4806	60835	24/9/1938	*The Green Howard, Alexandra, Princess of Wales's Own Yorkshire Regiment*
4818	60847	3/4/1939	*St. Peter's School, York A.D. 627*
4831	60860	15/6/1939	*Durham School*
4843	60872	20/5/1939	*King's Own Yorkshire Light Infantry*
4844	60873	20/6/1939	*Coldstreamer*
3676	60964	29/4/1958	*The Durham Light Infantry*

The very last V2 to remain in service was No 60836. From September 1966, it was the sole survivor of the class but it was retired from Dundee shed in December that year. Throughout most of its final year, No 60836 was very active and this picture of it at Dundee in September 1966 confirms that it was far from a candidate for storage.

Photo: Keith Lawrence

diesels in Scotland, the first Scottish V2 was not withdrawn until March 1962 and, although eighteen were retired by the end of that year, the high failure rate of the Type 2 diesels required the retention of the remaining Scottish-based V2s at least for the time being. In their last years, the Scottish V2s worked on everything from the Aberdonian to trains of empty oil tanks and the last, Dundee-based No 60836, survived until the beginning of December 1966.

By the time the Scottish V2s were feeling the effects of dieselisation, their English counterparts were quite familiar with the same story. As diesel traction took over more and more duties on the East Coast main line, the activities of the V2s became increasingly restricted and, as if to emphasise the impending redundancy of steam, June 1963 was scheduled as the date for the cessation of steam haulage south of Peterborough and, significantly, the closure of Kings Cross shed. Not to be ignored, the V2s raised two mechanical digits to the rule-book and carried on working into Kings Cross until December. Between Peterborough and York, V2s

continued to be used regularly on parcels trains almost until the class became extinct with the withdrawal of No 60831 from York on 6 December 1966. At the time of retirement, the ages of the V2s ranged from eighteen to twenty-nine years old and, while a few of those with longer life-spans clocked up marginally over one million miles, a number of the shorter-lived engines went to their graves with less than 700,000 miles on their clocks. For locomotives which incorporated some of the best ideas of one of the most celebrated designers in British locomotive history, those mileage figures represented little more than a trip round the block.

The V2s were the last new type of express locomotives designed by Nigel Gresley and it was often argued that they were his best all-round machines. Long before the class faced extinction, it had been decided that it would be appropriate to preserve an example of this significant design and, when No 60800 *Green Arrow* was withdrawn from Kings Cross shed on 21 August 1962, it was the obvious candidate. Surprisingly, none of the other V2s were acquired by preservation organisations and so the

TABLE 6.8: YEARLY TOTALS OF V2 CLASS 2-6-2s.

Totals taken at 31 December each year.

1936	5	1941	134	1946	184	1951	184	1955	184	1959	184	1963	72
1937	25	1942	163	1947	184	1952	184	1956	184	1960	184	1964	40
1938	44	1943	179	1948	184	1953	184	1957	184	1961	184	1965	14
1939	86	1944	184	1949	184	1954	184	1958	184	1962	115	1966	0
1940	118	1945	184	1950	184									

Only one V2 was preserved but, fittingly, it was the class leader *Green Arrow*. This excellent picture shows it in full LNER green livery, complete with original number, leaving Chester on 9 September 1991 with the North Wales Coast Express.
Photo: P. J. Chancellor

decision to set aside No 60800 was most fortunate. The restoration of *Green Arrow* which included the application of its original livery and number, 4771, was finished in Doncaster in October 1964 but, for the next seven and a half years, the locomotive was transferred from one storage site to another so frequently that it saw parts of Britain which it had not visited during its twenty-six year working life.

Eventually, it was realised that No 4771 would be quite happy at the National Railway Museum and, prior to being despatched there, it was res-

tored to working order at Norwich shed. Its first trip after being steamed again in March 1973 was from Norwich to Ely and, apart from the significance of seeing the locomotive in action once again, the trip marked the first appearance of a V2 on ex-Great Eastern tracks east of March for almost twenty years. Since then, No 4771 *Green Arrow* has put in countless appearances on main line specials and also as a guest at preservation centres throughout the country.

CHAPTER 7

Sir Nigel's Final Legacies

With the building of the A3 4-6-2s and, later, the streamlined A4s, Nigel Gresley had provided the LNER with an awesome array of express passenger locomotives. The V2 2-6-2s did for fast freight traffic what the 4-6-2s had done for passenger services, while the O2 2-8-0s supplied the muscle for the heaviest goods turns. Many suburban services were handled admirably by the N2 and N7 0-6-2Ts and the V1 2-6-2Ts whereas general freight and shunting duties benefited from, in particular, the attentions of J39 0-6-0s and 0-6-0Ts of the J50 and J52 classes.

One of the very few gaps that remained in the LNER's range of motive power was rather specialised in nature but, much to the exasperation of Edward Thompson, Gresley had previously demonstrated that he thought it quite practical to design a completely new type of locomotive to fulfil a specific role. The gap which required filling was, in the best Gresley tradition, to receive the master's full attention. By the mid-1930s, it was realised that stronger steeds were required for the West Highland line, the former North British Railway route between Glasgow and Mallaig, which had provided many a sleepless night for a succession of locomotive engineers. When the line had first been built, the terrain through which it passed had made it impossible to avoid steep gradients and sharp curves and, right from the beginning, the North British struggled to supply locomotives which combined short wheelbases and light axle-loadings with adequate power.

Although the West Highland line passed through vast areas which were devoid of population, it was well used. The locals in and around Fort William looked on the railway as the only sensible means of transport to the great beyond, particularly when compared with a road crossing of Rannoch Moor in winter, and the fishermen of Mallaig came to depend on the line for transporting their catches to markets inland. During the holiday season, the tourist trade to the West Coast of Scotland meant that the trains to Fort William and Mallaig were packed, and even the Glen class superheated 4-4-0s, which the North British had built for the line between 1913 and 1920, had to work double-headed on trains in excess of five coaches. If the construction of the line had required major feats of civil engineering, the working of the line could almost guarantee premature alopecia among the operating staff.

The LNER would have liked to use Gresley's K3 2-6-0s on the line in order to cut down on the double-heading but they were considered too heavy and, instead, thirteen of the lighter K2s were transferred to the West Highland line in 1924 and 1925. The loadings allowed for the K2s were 220 tons as opposed to the 180 tons permitted for the Glens and, for many years, the K2s became synonymous with the run to Mallaig. In 1933, the thirteen West Highland K2s were well and truly 'localised' by being named after Scottish lochs, although many Highlanders felt that the LNER was playing safe by ignoring any names which hinted of Gaelic origins.

In the 1930s, constant increases in the loadings of trains on the West Highland line began to place a strain even on the K2s and Gresley's solution to the motive power problem was the K4 class 2-6-0s. At first, authorisation was given for the construction of just one K4 and, although design work for

The first of Gresley's deceptively powerful K4 2-6-0s appeared in 1936 but it was 1938 before any of the other five entered service. The first of the 1938 batch was LNER No 3442 *MacCailein Mor* but it was soon renamed with the now-familiar handle of *The Great Marquess*. After withdrawal at the end of 1961, the locomotive was preserved and restored to its original condition and it is this guise in which it was photographed at Nine Elms in 1967.

Photo: Rail Archive Stephenson

TABLE 7.1: ORIGINAL DIMENSIONS OF K4 CLASS 2-6-0s

BUILT:	Darlington 1937/1938
WEIGHTS FULL (locomotive):	68 tons 8 cwt
(tender):	44 tons 4 cwt
TENDER CAPACITY (water):	3,500 gallons
(coal):	5 tons 10 cwt
WHEELBASE (locomotive):	25ft 2in
(tender):	13ft 0in
WHEEL DIAMETERS (leading):	3ft 2in
(coupled):	5ft 2in
CYLINDERS:	(3) 18½in × 26in
BOILER PRESSURE:	200 lb (a)
HEATING SURFACES (tubes):	871.1 sq ft
(flues):	382.5 sq ft
(firebox):	168.0 sq ft
(superheater):	310 sq ft
SUPERHEATER:	Robinson 24-element
GRATE AREA:	27.5 sq ft
TRACTIVE EFFORT @ 85% boiler pressure:	36,599 lb (a)
LNER ROUTE AVAILABILITY:	6
BR POWER CLASSIFICATION:	6MT

Note:

(a) Until June 1937, LNER No 3441 (later BR No 61993) had a boiler pressure of 180 lb and its tractive effort was 32,939 lb.

the new locomotive did not start until May 1936, the completed machine emerged from Darlington works to enter traffic in January 1937. Gresley incorporated as many tried and tested components as possible in the design and the result was, technically, a bit of a hybrid as the 18½in × 26in cylinders, the motion and the pony truck were based firmly on those of the K3s, the boiler was a slightly shortened version of those used by the K2s while the firebox design was borrowed from the B17 4-6-0s. The driving wheel diameter of 5ft 2in was, however, something new for a mixed traffic locomotive.

The solitary K4 carried No 3441 and, in order to continue the naming theme of the West Highland K2s, was named *Loch Long*. Wearing the standard LNER mixed traffic livery of lined black, it was despatched to Eastfield shed in Glasgow and spent its first five weeks on general freight duties so that the crews could familiarise themselves with it. On 4 March 1937, it made its debut on the West Highland line and, within little over a month, was making a daily trip from Glasgow (Queen Street) to Fort William and back on a passenger service. The K4 quickly became the subject of much jealousy and possessiveness among the crews at Eastfield and Fort William sheds, as its ability to haul 300-ton trains without assistance represented an improvement of almost 50% on the K2s. Furthermore, it was found to use no more water than the K2s consumed when they were hauling loads of just 200 tons, and so a single water stop at Crainlarich sufficed for the K4 while a struggling K2 had often had to take on extra water at Bridge of Orchy.

Despite its strength, No 3441 was found to be little faster than the older locomotives on long uphill climbs. By the time this drawback was fully appreciated, the holiday season was drawing near and the operating department did not relish the embarrassment of the new locomotive struggling with the anticipated peak summer loadings. In order to rectify the locomotive's sluggishness on hills, it was sent to Cowlairs works in June 1937 to have its boiler pressure raised from 180lb to 200lb, and the slight loss of adhesion which resulted was more than compensated for by an increase in tractive effort from 32,939lb to 36,599lb, over 1,000lb greater than the A4s. The modification

worked a treat.

One other problem was not solved. This was the locomotive's tendency to give a rough ride at high speeds and, curiously, this seemed to be an inherent feature of all Gresley's 2-6-0s. The section of the West Highland line between Glasgow and Craigendoran was quite level and, as speeds of 60 mph were commonplace on that stretch, footplate crews could take a bit of a battering but the sprint to Craigendoran accounted for only 22½ miles of the journey to Fort William. The 100-mile section north of Craigendoran had the little matter of Corrour summit which was 1,347 feet above sea level and, on the southbound journey, was reached by six miles of unbroken 1 in 67. On the northbound run, Ardlui summit was reached by a fifteen-mile climb of around 1 in 60. Any crew members who complained of rough riding at 60 mph over those obstacles would have been considered well and truly out to lunch.

Throughout the summer of 1937, No 3441 did not grumble about its loadings or the intensity of its scheduling. The locomotive's ability to complete the arduous 245-mile round trip between Glasgow and Fort William in one day was something that the K2s were only rarely burdened with and, for the staff at Eastfield, this was akin to having two additional engines and not just one. Gresley had no hesitation in ordering five more K4s early the following year although only one, No 3442, was completed in time for the heaviest summer traffic on the West Highland line. No 3442 arrived at Eastfield in July 1938 and, although that was just in time for the peak holiday season, it was put to work initially on general freight turns which only occasionally involved runs to Fort William. It was not until 5 August that No 3442 made its first appearance on a passenger working to Fort William. The four other K4s, Nos 3443–46, did not emerge from Darlington until December 1938 and it was the following month before the last two of those officially entered traffic.

In contrast to the class leader, Nos 3443–46 were turned out in fully-lined apple green liveries and, although all were named, the theme of lochs was not continued. The name of *MacCailein Mor* was applied to No 3443 at first but, within a few weeks, this was replaced by *The Great Marquess*.

131

Nos 3443/44 were named *Cameron of Lochiel* and *Lord of the Isles* respectively while No 3445 carried *MacCailin Mor*, the amended version of No 3443's original name. The final K4, No 3446, was named *Lord of Dunvegan* but it took a member of the Dunvegan family to point out to the LNER that the last Lord had passed away in 1935 and had left no male heir. Hiding its red corporate face, the LNER diplomatically renamed No 3446 *MacLeod of MacLeod* in March 1939.

Clan loyalty in Scotland is evident to this very day, and the early renaming of No 3443 from *MacCailein Mor* to *The Great Marquess* suggested to the Highlanders that some mere Englishman on the LNER had not done his homework. MacCailin Mor, to use the correct spelling, was the familiar Gaelic name for the Duke of Argyll who was the chief of the Clan Campbell, while The Great Marquess referred to James Graham, the Marquess of Montrose. In 1645, the hostilities between the Campbells and the Grahams had reached such a bloody peak that the Scottish population had been reduced by some

three thousand. Although the Campbells came off the worst by far, there had been little sympathy in West Highland territory as the Marquess of Montrose lived at Inverlochy Castle, just a stone's throw from Fort William and so the Grahams had been considered, most definitely, the home side. As if to compound its treachery, the LNER seemed not to have consulted the railway history books as, when the West Highland line was being planned in the 1880s and 1890s, one of its strongest opponents happened to be the Duke of Argyll, none other than the reigning MacCailin Mor himself.

When No 3443 was renamed *The Great Marquess*, it appeared that the LNER had acknowledged its errors, but the reappearance of the name *MacCailin Mor* on No 3445 confounded the locals. It was said that it was easy to tell which staff members at Eastfield and Fort William were Campbells and which were Grahams by the positioning of No 3443 and No 3445 at the sheds. If either of the two locomotives was given pole position it was the result of allied collaboration but, wherever one was

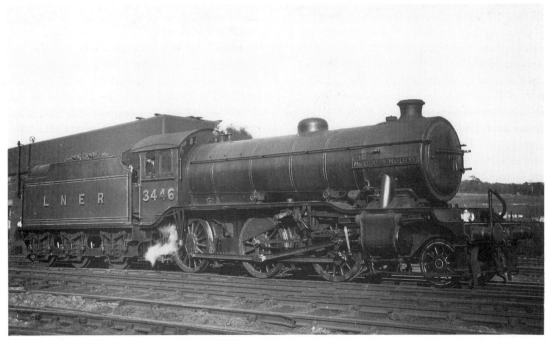

The last of the six K4s was LNER No 3446 which did not officially enter traffic until January 1939; originally named *Lord of Dunvegan*, its name was changed after just three months to *MacLeod of MacLeod*. Although undated, this picture must have been taken in 1939 or 1940 as No 3446 and its classmates succumbed to the obligatory black livery during the war. It later became BR No 61998 and survived until October 1961.

Photo: H. C. Casserley Collection

In 1955, K4 No 61994 *The Great Marquess* was not kept quite as spotless as it had once been. Here, it is on shed at Eastfield being prepared for its first turn of the day.   *Photo: D. K. Jones Collection*

parked, the other would be kept at the maximum possible distance!

The section of the West Highland line between Fort William and Mallaig was not as difficult a route as that between Glasgow and Fort William, and so the older K2s could usually cope with duties beyond Fort William without the need for double-heading. This left the more-powerful K4s available for the challenge of the haul over Rannoch Moor and, for almost ten years, it was extremely rare to see a K4 sauntering off on a jaunt to Mallaig.

The K4s had been designed for one specific route and, as six were considered sufficient for at least the foreseeable future, no more were ever built. Apart from the transfer of Nos 3443/44 from Eastfield to Fort William in October 1939 and the application of the obligatory black livery to Nos 3442–46 after the outbreak of war, life on the West Highland went on almost unchanged for the K4s until December 1945. In that month, Gresley's successor, Edward Thompson, rebuilt No 3445 with a new 225lb boiler and substituted two cylinders for its original three.

The intention was to use the rebuild as the prototype for a new class of freight engine and it was tested at various depots throughout the LNER system from New England to Thornton Junction, but it took until July 1947 for the order to be placed for construction of similar locomotives. The rebuilt K4 and the new locomotives were designated K1, that classification having become vacant in 1937 as a result of the last of the original ex-Great Northern K1s having been converted to K2s. By the time the first purpose-built K1 appeared in May 1949, not only had Edward Thompson retired from the LNER but his successor, Arthur Peppercorn, had seen his incumbency terminated abruptly by Nationalisation.

Despite its reluctant membership of the K1 class, No 3445 remained in the numbering sequence of the unadulterated K4s. Thompson's renumbering scheme of 1943 had allocated Nos 1993–98 to the K4s and, when the new numbers were eventually applied in 1946, No 3445 became No 1997 although it had, by then, been reclassified as K1. The K4s were not affected by the LNER's renumbering of 1946 and so they passed to British Railways to

142 **Tables 98, 99**

Table 98 Edinburgh and Glasgow to Fort-William and Mallaig

	Sats only p.m (Sleeping Car (1st and 3rd Class) London to Fort William; TC London to Fort William; Begins 13th June)	p.m (Through and Sleeping Car (1st and 3rd Class) London to Fort-William — Sunday to Thursday nights inclusive also Friday 4th June; RC Glasgow to Fort-William)	a.m.	p.m (RC Glasgow to Fort William; Begins 13th June)	p.m	Sats only a.m	.	p.m	a.m (RC Glasgow to Fort-William; Through Train Glasgow to Mallaig)	Ex Sats p.m	Sats only p.m	.
London (King's X) lev	7 F 0	7 E 0	—	10 0E	1030E	1 0	—	—	1 0	—	—	—
Edinburgh (Waverley) lev.	4a20	4a20	6 25	9 a 5	9 a 5	11 0	.	.	2 p 0	4 0	4 0	.
Glasgow Queen St. (Low Level)	.	—	7B50	.	1032B	1B10	—	—	—	5B30	6B7	—
Glasgow Queen St. (High Level)	5 5	5 55	.	1021	.	.	.	.	3 46	.	.	.
Dumbarton	—	6 22	8B24	1050	11B3	1B38	—	—	3B48	5B55	6B36	—
Craigendoran	.	.	9 0	11 3	1130	2 0	.	.	4 22	6 9	7 0	.
Helensburgh (Upper)	—	6 39	9 6	1110	1136	2 6	—	—	4 29	6 15	7 6	—
Rhu	.	.	9 11	—	1141	2 11	.	.	4 34	6 20	7 11	.
Shandon	.	—	9 17	.	1147	2 17	—	—	4 40	6 29	7 18	—
Garelochhead	6 18	.	9 25	—	1155	2 25	.	.	4 46	6 39	7 28	.
Whistlefield	.	—	9 30	.	12 0	2 30	—	—	4 52	6 44	7 33	—
Arrochar and Tarbet	.	7 17	9 51	1152	1221	2 51	.	.	5 16	7 7	7 54	.
Ardlui (Loch Lomond)	—	7 34	—	1213	—	—	—	—	5 38	—	—	—
Crianlarich arr	7 16	7 55	.	1234	.	.	.	.	6 0	.	.	.
Crianlarich lev.	7 23	8 1	—	1240	—	—	—	—	6 6	—	—	—
Tyndrum	.	8 14	.	1253	.	.	.	.	6 20	.	.	.
Bridge of Orchy	—	8 29	—	1 8	—	—	—	—	6 35	—	—	—
Rannoch (for Kinloch-Rannoch)	.	8 58	.	—	.	.	.	.	7 2	.	.	.
Corrour	—	9 14	—	.	—	—	—	—	7 18	—	—	—
Tulloch (Loch Laggan and Kingussie)	.	9 35	.	2 11	.	.	.	.	7 39	.	.	.
Roy Bridge	—	9 47	—	2 23	—	—	—	—	7 51	—	—	—
Spean Bridge	.	9 57	.	2N38	.	.	.	.	8 0	.	.	.
Fort-William A arr.	9 30	1014	—	3 4	—	—	—	—	8 17	—	—	—
Fort-William A lev.	9 50	1028	.	.	.	.	.	4 50	8K30	.	.	.
Banavie	—	1035	—		—	—	—	4 57	8K37	—	—	—
Corpach	.	1039	.		.	.	.	5 1	8K11	.	.	.
Locheilside	—	1052	—		—	—	—	5 13	8K53	—	—	—
Glenfinnan M (Loch Shiel)	1022	11 6	.		.	.	.	5 27	9K7	.	.	.
Lochailort	1044	1127	—		—	—	—	5 48	9K28	—	—	—
Arisaig	11 3	1146	.		.	.	.	6 7	9K17	.	.	.
Morar	1116	1159	—	—	—	—	—	6 21	10K1	—	—	—
Mallaig arr.	1123	12 6	.	.	.	.	.	6 28	10K8	.	.	.

* Saturdays only † Except Sats. § Mondays only ‡ Except Mondays
A Station for Ballachulish (Glencoe) and Kinlochleven B Change Craigendoran
E Sunday to Friday nights inclusive F Friday nights only K Thursdays 5 minutes later
M Connection from Glenfinnan Pier to Dalilea and Acharacle off 5-55 a.m. Glasgow to Mallaig
N Arrive 2-30 p.m RC Restaurant Car TC Through Carriage

For Steamer Sailings to Portree and Stornoway, apply D. MacBrayne Ltd. 44 Robertson St., Glasgow C.2.

Table 99 Edinburgh and Glasgow to Dumbarton, Helensburgh and Balloch
Sundays

Leave	A a.m	A a.m.	a.m	.	A a.m	n'n	A a.m.	.	A p.m	a.m	.	A p.m	p.m	.	A p.m	A p.m	p.m	.	A p.m	.
Edinburgh Waverley Stn.	7 40	—	7 40	—	—		11 0	—	—	11 0	—	—	—	—	—	—		—	6 30	—
Bridgeton X lev.	1020	1030	1040		1130	12 0	1 15	.	1 25	2 5	.	2 10	3 15	.	5 55	7 5	7 15	.	8 20	.
Queen St. (Low)	1030	1040	1050	—	1140	12 0	1 25	—	1 35	2 15	—	2 50	3 25	—	6 5	7 15	7 25	—	8 30	—
Partick	1038	1048	1058	.	1148	1218	1 33	.	1 43	2 23	.	2 58	3 33	.	6 13	7 23	7 33	.	8 38	.
Drumchapel	1046	—	11 6	—	1156	—	1 41	—	—	2 31	—	—	—	—	—	—	—	—	—	—
Singer	1050	1058	1110	.	12 0	1228	1 45	.	1 53	2 35	.	3 8	3 43	.	6 23	7 33	7 43	.	8 48	.
Dumbarton	11 1	11 9	1121	—	1211	1239	1 56	—	2 4	2 46	—	3 19	3 54	—	6 34	7 44	7 54	—	8 59	—
Cardross	11 8	1116	.	.	1218	.	2 3	.	2 11	.	.	3 26	.	.	6 41	7 51	.	.	9 6	.
Craigendoran	1114	1122	—	—	1224	—	2 9	—	2 17	—	—	3 32	—	—	6 47	7 57	—	—	9 12	—
Helensburgh arr.	1120	1128	.	.	1230	.	2 15	.	2 23	.	.	3 37	.	.	6 52	8 2	.	.	9 17	.
Alexandria	—	—	1127	—	—	1245	—	—	—	2 52	—	—	4 0	—	—	—	8 0	—	—	—
Balloch arr.	.	.	1121	.	.	1249	.	.	.	2 56	.	.	4 4	.	.	.	8 4	.	.	.
Balloch Pier "	—	—	1137	—	—	1253	—	—	—	3 2	—	—	4 8	—	—	—	8 8	—	—	—

A Ceases after 5th September

Tables 98—101 143

Mallaig and Fort-William to Glasgow and Edinburgh — Table 98

	a.m	a.m	.	a.m	a.m (RC Fort-William to Glasgow)	Sats only p.m	Ex Sats p.m	Sats only p.m	Sats only p.m (Begins 13th June)	p.m (Through and Sleeping Cars (1st and 3rd Class) Fort-William to London (arrive 7-15 a.m); RC Fort-William to Glasgow)	(RC Fort-William to Glasgow; Begins 13th June)	p.m (Except Thursdays until 9th September daily thereafter)	p.m (Thursdays only and ceases after 9th September)
Mallaig lev.	—	6 35	—	—	7 46	—	—	—	1230	1 0		5 42	6 40
Morar	.	6 46	.	.	7 55	.	.	.	.	1 9		5 51	6 49
Arisaig	—	7 1	—	—	8 8	—	—	—	—	1 22		6 8	7 6
Lochailort	.	7 24	.	.	8 27	.	.	.	.	1 41		6 27	7 25
Glenfinnan M (Loch Shiel)	—	7 50	—	—	8 52	—	—	—	1 27	2 2		6 48	7 46
Locheilside	.	8 11	.	.	.	.	.	.	.	2 17		.	.
Corpach	—	8 30	—	—	—	—	—	—	—	2 30		7 16	8 14
Banavie	.	8 37	.	.	.	.	.	.	.	2 35		7 21	8 19
Fort-William A arr.	—	8 46	—	—	9 24	—	—	—	1 59	2 42		7 28	8 26
Fort-William A lev.	.	.	.	.	9 36	.	.	.	2 15	2 55	4 0		
Spean Bridge	—	—	—	—	9 56	—	—	—	—	3 14	4 18		
Roy Bridge	.	.	.	.	10 3	.	.	.	.	3 21	4 25		
Tulloch (Loch Laggan & Kingussie)	—	—	—	—	1021	—	—	—	3 0	3 41	4 42		
Corrour	.	.	.	.	1043	.	.	.	.	4 4	.		
Rannoch (for Kinloch-Rannoch)	—	—	—	—	1057	—	—	—	—	4 18	—		
Bridge of Orchy	.	.	.	.	1122	.	.	.	.	4 46	.		
Tyndrum	—	—	—	—	1139	—	—	—	—	5 4	5 56		
Crianlarich arr.	.	.	.	.	1148	.	.	.	4 24	5 13	6 5		
Crianlarich lev.	—	—	—	—	1154	—	—	—	4 34	5 19	6 11		
Ardlui (Loch Lomond)	.	.	.	.	1212	.	.	.	.	5 37	.		
Arrochar and Tarbet	7 18	—	—	10 6	1231	1 0	2 45	4 50	5 17	5 55	6 46		
Whistlefield	7 42	.	.	1028	.	1 21	3 8	5 15		6 17	.		
Garelochhead	7 47	—	—	1034	1257	1 28	3 15	5 20		6 22	—		
Shandon	7 53	.	.	1041	1 4	1 34	3 22	5 26		6 28	.		
Rhu	7 59	—	—	1047	—	1 40	3 29	5 32		—	—		
Helensburgh (Upper)	8 4	.	.	1052	1 15	1 45	3 35	5 37		6 40	7 29		
Craigendoran arr.	8 8	—	—	1056	1 19	1 49	3 40	5 41	6 3	6 44	—		
Dumbarton arr.	8E27	.	E	1120	1E37	2E21	4E10	6E0	6 18	7E6R	7 46	.	.
Glasgow Queen Street High Level	—	—	—	—	2F4	—	—	—	—	7L30	8S26	—	—
Glasgow Queen Street Low Level	8E59	.	E	1154	.	3E3	4E46	6E41	6 46	.	.	.	.
Edinburgh (Waverley) arr.	1112	—	—	1B25	3 43	4Q45	6 16	9 2	9 2	9 24	1030	—	—
London (King's X) arr.	10p5	K		10N5	P4a50	4a50	6a5	6a20	6a20	6a30	.	.	.

For Steamer Sailings from Stornoway, Portree, and Inverness in connection with Trains, apply D. MacBrayne Ltd., 44 Robertson Street, Glasgow C.2

* Sat only † Ex Sat § Mon only ‡ Ex Mon
A Station for Ballachulish (Glencoe) and Kinlochleven B On Fris. arrive 2-9 p.m and on Sats. 1-33 p.m E Change Craigendoran
F Stops Cowlairs 1-55 p.m
K Fridays and Saturdays 9-55 p.m. Passengers can arrive London from 5th July at 8-10 p.m by Pullman Car Train at Supplementary Charges.
L Stops Cowlairs 7-21 p.m
M Connection from Acharacle and Dalilea to Glenfinnan Pier into 1-0 p.m Mallaig to Glasgow
N 2-55 a.m on Saturday and Sunday mornings
P Saturdays 4-40 a.m. Q 5-4 p.m. on 5th June
R From 13th September arrive 7-17 p.m.
S Stops Cowlairs 8-17 p.m.
RC Restaurant Car TC Through Carriage

Balloch, Helensburgh & Dumbarton to Glasgow & Edinburgh — Table 99
Sundays

Leave	p.m	A p.m	A p.m	.	p.m	A p.m	.	p.m	A p.m	.	p.m	A p.m	.	A p.m	A p.m	A p.m	.	p.m	A p.m	.
Balloch Pier	1210	—	—	—	3 20	—	—	5 5	—	—	6 55	—	—	—	—	—	—	8 40	—	—
Balloch Station	1215	.	.	.	3 25	.	.	5 10	.	.	7 0	.	.	.	.	.	.	8 45	.	.
Alexandria	1220	—	—	—	3 30	—	—	5 15	—	—	7 5	—	—	—	—	—	—	8 50	—	—
Helensburgh	.	1210	2 18	.	.	4 0	.	.	5 45	.	.	.	.	7 10	.	8 25	.	.	9 35	.
Craigendoran	—	1215	2 28	—	—	4 5	—	—	5 50	—	—	7 5	—	7 15	8 22	8 28	—	—	9 45	—
Cardross	.	1224	2 36	.	.	4 14	.	.	5 59	.	.	7 13	.	7 24	8 30	8 36	.	.	9 53	.
Dumbarton	1227	1233	2 45	—	3 38	4 23	—	5 22	6 8	—	7 12	7 22	—	7 33	[illegible]	8 45	—	8 57	10 2	—
Singer	1238	1244	2 56	.	3 49	4 34	.	5 33	6 19	.	7 23	7 33	.	7 44	8 50	8 56	.	9 8	1013	.
Drumchapel	—	—	—	—	3 53	—	—	—	—	—	7 27	7 37	—	—	—	9 0	—	—	1017	—
Partick	1248	1254	3 6	.	4 1	4 44	.	5 43	6 29	.	7 35	7 45	.	7 54	9 0	9 8	.	9 18	1025	.
Queen St. (Low)	1255	1 1	3 13	—	4 8	4 51	—	5 50	6 36	—	7 42	7 52	—	8 1	9 7	9 15	—	9 25	1032	—
Bridgeton Cross	1 2	1 8	3 20	.	4 15	4 58	.	5 57	6 43	.	7 49	7 59	.	8 8	9 14	9 22	.	9 32	1039	.
Edinburgh (Wav)	—	3 23	..	—	—	7 18	—	7 18		—	9 18	9 18	—	1038	—	—	—	—	—	—

A Ceases after 5th September

Spean Bridge, Invergarry and Fort Augustus. Table 100

The Train Service has been withdrawn from Gairlochy, Invergloy, Invergarry, Aberchalder, and Fort Augustus, but Parcels and Miscellaneous Passenger Train traffic previously dealt with at these stations will continue to be accepted. For Motor Service see Table 224.

Ferry Service—Mallaig and Armadale (Isle of Skye) Table 101
Bus Service—Armadale, Broadford, Sligachan and Portree

A twin screw ferry boat (privately owned) seating 120 passengers will, weather permitting, leave Mallaig for Armadale on arrival of trains from South, Edinburgh and Glasgow, due Mallaig 12-6 p.m., arriving Portree 3-15 p.m. Bus leaves Portree 8-45 a.m. to connect with ferry and train leaving Mallaig 1-0 p.m for Glasgow, Edinburgh and the South. Bus also leaves Portree 2-30 p.m to connect with ferry and train leaving Mallaig 5-42 p.m (6-40 p.m on Thursdays until 9th September) for Fort-William.

Services over the West Highland line in the first year of nationalisation are detailed in the Scottish Region's timetable for the summer of 1948. The K4 2-6-0s dominated the services at this time.

become Nos 61993–98, with the rebuilt locomotive defiantly carrying No 61997. Ironically, No 61997 was transferred to Eastfield shed in 1949 and, five years later, was reallocated to Fort William. Like Nos 61993/94, the rebuild never managed to shrug off a black livery but, although Nos 61995/96/98 were eventually restored to apple green, they later succumbed to black as well.

When No 61997 was transferred to Fort William shed in 1954, it did not rejoin its former pals, as the depot's two K4s were reallocated simultaneously to Eastfield. The movements were not done to prevent a conspiracy between *MacCailin Mor, Cameron of Lochiel* and *Lord of the Isles*, but came about as a result in the change of motive power on the West Highland line. In 1947, the B1 class 4-6-0s had started to appear on the line and the K4s were subsequently relegated to either freight turns or duties on the Mallaig section. The allocation of several of the new K1s to the West Highland line in 1949 was followed, one year later, by the introduction of Black Fives. The use of the characterless ex-LMS creatures was a knock-on effect of administrative reorganisation in Scotland during which

Fort William depot became a sub-shed of Perth and, as Perth had an abundance of the 4-6-0s, it was quite happy to let a few of them loose on the West Highland line.

The arrival of the new types of locomotives restricted the usefulness of the K4s even more and, from 1954, their duties on the line consisted almost entirely of freight workings with the very occasional passenger turn between Fort William and Mallaig thrown in. The Eastfield shed staff were not averse to finding the K4s work elsewhere and, during the 1950s, the locomotives were known to work to Edinburgh, Perth, Forfar, Ayr and even Tweedmouth.

The K4s' long association with the West Highland line ended, in theory at least, in 1959 when they were transferred to Thornton Junction shed for general freight duties. Their flag was kept fluttering by No 61997 which, in its K1 cloak, remained at Fort William until withdrawal in June 1961. The end for the five original K4s came soon afterwards with four being retired in October 1961 and the remaining one, No 61994, being taken out of service just two months later. After their transfer

Between 1954 and 1959, all five unrebuilt K4s were shedded at Eastfield. This picture shows No 61998 *MacLeod of MacLeod* taking on water at the shed on 26 August 1955.

Photo: E. H. Sawford

K4 No 61995 *Cameron of Lochiel* was transferred from Eastfield to Thornton Junction shed in December 1959 but it was to make one popular return visit to its former haunt on 18 June 1960. On that date, it hauled an SLS special from Glasgow to Fort William but, as can be seen from the photograph, the combination of a fully-stocked bar and a lack of toilet facilities necessitated regular stops across Rannoch Moor.

Photo: Rail Archive Stephenson

K4 class 2-6-0 No 61995 *Cameron of Lochiel* produces a good head of steam as it leaves Fort William on 31 July 1952 with the 4.40pm to Glasgow.

Photo: E. R. Morten

to Thornton Junction in 1959, the only known working of a K4 over the West Highland line was on 18 June 1960 when No 61995 *Cameron of Lochiel* hauled a Stephenson Locomotive Society special train from Glasgow to Fort William.

The withdrawal of No 61994 *The Great Marquess* did not condemn the K4s to oblivion as the locomotive was purchased by Viscount Garnock and restored to full working order, complete with LNER livery and No 3443, at Cowlairs works. It was returned to action on 29 April 1963 and went on to see regular use on enthusiast's specials before being acquired by the Severn Valley Railway in 1972.

Nigel Gresley's last new design was for the V4 class 2-6-2s. The idea of a new lightweight 2-6-2 for branch and secondary duties had been tossed around the LNER's corridors since 1939 and Gresley had seen, at a distance, the advantages of light six-coupled locomotives elsewhere. Although he was not a disciple of standardisation, he had to admit that William Stanier's dull and dreary Black Fives had proved immensely useful for the LMS while, down in the West Country, Charles Collett's Hall and, more particularly, Grange class 4-6-0s had added a new dimension to the Great Western's motive power stock.

The LNER's nearest equivalent to the LMS and GWR 4-6-0s were the K2 and K3 class 2-6-0s. The former type was, by the late 1930s, a little outdated, whereas the latter's axle weight of over twenty tons made it unsuitable for many of the LNER's secondary lines. After the idea for new 2-6-2s was raised in 1939, the first drawings were for a locomotive with a boiler pressure of 300lb and three cylinders of 14in × 26in but later drawings incorporated a boiler pressure of 250lb and, in order to partially offset the theoretical loss of power, the diameter of the cylinders was increased to 15in. One factor which was retained throughout all the stages of the design work was the driving wheel diameter of 5ft 8in.

A vital requirement was that the weight of the new locomotive should be kept as low as possible, and so it was proposed to use a proportion of nickel steel in the construction of the boiler and coupling rods and also to use welded and fabricated components instead of castings as far as possible. It was estimated that those two measures alone would save some three and a half tons in weight. The target axle weight for the new design was no more than just seventeen tons in order to maximise its route availability and eventually this was achieved.

The order for the two new locomotives was placed in October 1939 but it mattered little that the design details had not, by then, been finalised. The outbreak of war meant that a severe shortage of materials and manpower was just around the corner and so the construction of the engines would not happen overnight. As things turned out, it was February 1941 before the first V4 emerged from Doncaster works while the second one followed a

TABLE 7.2: THE K4 CLASS 2-6-0s

FIRST NO.	1946 NO.	BR NO.	NAME	BUILT	SHED ALLOCATIONS		WDN.
					1/1/1948	1/1/1960	
3441	1993	61993	*Loch Long*	1/37	Eastfield	Thornton Jct	10/61
3442	1994	61994	*The Great Marquess**	7/38	Eastfield	Thornton Jct	12/61
3443	1995	61995	*Cameron of Lochiel*	12/38	Fort William	Thornton Jct	10/61
3444	1996	61996	*Lord of the Isles*	12/38	Fort William	Thornton Jct	10/61
3445	1997†	61997†	*MacCailin Mor*	12/38	New England	Fort William	6/61†
3446	1998	61998	*MacLeod of MacLeod**	12/38	Eastfield	Thornton Jct	10/61

* Renamed Locomotives
LNER No 3442 (later BR No 619944) named *MacCailein Mor* until July 1938.
LNER No 3446 (later BR No 61998) named *Lord of Dunvegan* until March 1939.
† LNER No 3445 (later BR No 61997) rebuilt as K1 class December 1945.

month later. They were given Nos 3401/02 and, as they were completed a few months before the official decree of black liveries for everything, they were turned out in the LNER's customary apple green colour scheme. Externally, the two locomotives were identical but, internally, they had different fireboxes. Whereas No 3401 had a standard copper firebox, No 3402 had a welded steel firebox which was fitted with a Nicholson thermic syphon to improve the circulation of the water.

The first of the V4s, No 3401, was named *Bantam Cock* and although the second, No 3402, was destined to remain anonymous in the eyes of officialdom, crews and enthusiasts alike referred to it as 'Bantam Hen'. At first, both V4s were allocated to Doncaster and they used the route between Doncaster and Leeds for their obligatory trials. No 3401 was despatched to Haymarket shed in Edinburgh in May for further trials and, the following month, the staff at Kittybrewster shed in Aberdeen were allowed to play with it for three

weeks before its departure for Stratford shed. During its stay at Kittybrewster, one of its test routes was between Aberdeen and Elgin and few, if any, other LNER locomotives could have boasted that, during the same month, they hauled scheduled trains into both Elgin and Liverpool Street stations. Those two stations are over six hundred miles apart!

The peripatetic tendencies of No 3401 resurfaced in February 1942 when it was transferred to Haymarket shed once again and, as its duties sometimes took it to Glasgow, it occasionally met up with No 3402 which had been allocated to Eastfield depot the previous September. The opportunities for the pair of V4s to mull over old times did not often arise at first, as the usual turn of duty for No 3402 was helping out the K4s on the run from Glasgow to Fort William but, in October 1943, the staff at Haymarket shed took sympathy on the loneliness of No 3401 and transferred it to Eastfield to rejoin its old chum.

By the time the two V4s were reunited in

TABLE 7.3: ORIGINAL DIMENSIONS OF THE V4 CLASS 2-6-2s

BUILT:	Doncaster 1941
WEIGHTS FULL (locomotive):	70 tons 8 cwt
(tender):	42 tons 15 cwt
TENDER CAPACITY (water):	3,500 gallons
(coal):	6 tons
WHEELBASE (locomotive):	29ft 4in
(tender):	13ft 0in
WHEEL DIAMETERS (leading):	3ft 2in
(coupled):	5ft 8in
(trailing):	3ft 2in
CYLINDERS:	(3) 15in × 26in
BOILER PRESSURE:	250 lb
HEATING SURFACES (tubes):	884.3 sq ft
(flues):	408.2 sq ft
(firebox):	151.6 sq ft (a)
(superheater):	355.8 sq ft
SUPERHEATER:	22 elements
GRATE AREA:	28.5 sq ft
TRACTIVE EFFORT @ 85% boiler pressure:	27,420 lb
LNER ROUTE AVAILABILITY:	4
BR POWER CLASSIFICATION:	5MT (4MT from May 1953)

Note:

(a) No 3401 was fitted with a Nicholson thermic syphon which added 19.5 sq ft of heating surface to the firebox.

The first of the two V4 2-6-2s was completed in February 1941, just two months before the death of Sir Nigel Gresley. At first, the locomotive wore No 3401 but the LNER renumbering scheme of 1946 resulted in it wearing No 1700. It was named *Bantam Cock* and it posed for this picture at Fort William in 1947, just a year before becoming BR No 61700.

Photo: Rail Archive Stephenson

Glasgow, the LNER's motive power was in the care of Edward Thompson. During Gresley's reign, it had been anticipated that Nos 3401/02 would be the forerunners of a numerous class of locomotives, particularly as the two prototypes had proved to be very efficient engines. Thompson, however, was unenthusiastic about Gresley's pony truck design and the three-cylinder arrangement and, even though he was quietly impressed by some of the V4s' sophisticated engineering, because of the lowly nature of their intended duties, Thompson agreed with a colleague's opinion that they were the equivalent of Rolls-Royces doing Ford jobs and, to him, this smacked of extravagance.

The V4s were costly to build and maintain and, whereas Gresley had anticipated that these factors would be offset by the locomotives' fuel efficiency and extensive route availability, Thompson did not have the time, the funds, or the wish to see if expectations would become reality. Thompson preferred standardisation and, under the economic conditions of the war years, it was difficult to argue

against that policy. Almost as soon as his feet were under the table, he embarked on design work for his B1 4-6-0s and he had every intention of these locomotives becoming the LNER's all-purpose, do-anything, go-anywhere machines.

It did not seem to affect Nos 3401/02 when they realised that their family was not going to grow and, as the sole members of the V4 class, they carried on quite contentedly on the West Highland line for several years. It was inevitable that comparisons were made between the V4s and the K4s on the run between Glasgow and Fort William, and it was generally considered that there was little to choose between the two classes despite the fact that, on paper, the maximum permissible load of 250 tons for the V4s was 50 tons less than that of the K4s. Each type had its advantages and, in the case of the V4s, one of the main plus-points was that they were considerably smoother riders on the fast section between Glasgow and Craigendoran. The V4s' main disadvantage was that they were far more prone to wheel slip on greasy rails than the

The second, and last, of Gresley's V4 2-6-2s was unofficially christened *Bantam Hen*. It started life as LNER No 3402 in 1941 and ended its days as BR No 61701 in 1957, almost all of its time having been spent in Scotland. It was a very shy creature and only rarely posed for pictures although the sheer charm of photographer Eric Sawford managed to coax it into having its picture taken at Kittybrewster shed in Aberdeen on 24 August 1955.

Photo: E. H. Sawford

TABLE 7.4: THE V4 CLASS 2-6-2s

FIRST NO.	1946 NO.	BR NO.	SHED ALLOCATIONS				WDN.
			FIRST	1/1/42	1/1/48	1/1/57	
3401	1700	61700	Doncaster	Norwich	Eastfield	Ferryhill	3/57
3402	1701	61701	Doncaster	Eastfield	Eastfield	Ferryhill	11/57

K4s and, as rain was not exactly an uncommon occurrence in the West Highlands, the problem was not a rare one.

The LNER green liveries which the V4s had worn on their debuts in 1941 had been short lived and, at the time of Nationalisation, they were still in the black paintwork which they had received in the winter of 1942/43. They became British Railways Nos 61700/01 and, although No 61701 was restored to green livery in June 1948, No 61700 *Bantam Cock* did not enter the works for repainting until March 1949 and it swiftly discovered that, by then, lined black had been adopted for non-express engines. Later in 1949, No 61701 also received a black livery.

Shortly after Nationalisation, the Highlands were invaded by Thompson's rough-riding B1s and ex-LMS Black Fives and the V4s were, like the K4s, displaced from the Fort William line. Eastfield shed found the two V4s alternative work on everything from fish trains to double-headed expresses and similar duties awaited them when they were transferred to Ferryhill shed at Aberdeen in May 1954. Apart from just over a year when they were based at Kittybrewster shed for working over the former Great North of Scotland lines, Nos 61700/01 saw out their days at Ferryhill. During their stay at Kittybrewster, their axle weights of just seventeen tons had enabled them to explore parts of the old Great North system which were out of bounds for

many other medium-sized locomotives. When they were returned to Ferryhill, however, that shed had no turns which were off-limits for the B1s or Black Fives and, consequently, the V4s were withdrawn in 1957 despite the fact that they were just sixteen years old and each had less than half a million miles on its clock.

Although Edward Thompson had decided against continuing the V4 class, neither he nor Arthur Peppercorn made any attempt to rebuild the two examples. Apart from Thompson's quibbles about the V4s' pony trucks and cylinders, the general quality of the engineering work which went into the locomotives was seldom disputed and so there was no need for any major alterations. The policy of rebuilding small numbers of locomotives in order to standardise them as part of a larger class was accepted railway practice, but there was never any hint that this would be forced on the V4s.

It has often been suggested that the reason why the two V4s were left alone was out of respect for Sir Nigel Gresley. The first of the pair was completed in February 1941 and the second in March. On 5 April 1941, Gresley died and, although his V2 2-6-2s continued to be constructed after his death, the V4s were not only his very last new design but they had also gone the whole hog from conception to realisation while he was still alive. It was, therefore, quite inexcusable that neither of the two locomotives was saved for preservation.

Bibliography

Several publications have been consulted during the preparation of this book and they include:

Atlantic Era, Evans. Marshall.
BR Steam Motive Power Depots. Bolger. Ian Allan.
British Atlantic Locomotives. Allen. Ian Allan.
British Locomotive Catalogue 1825 – 1923. Baxter. Moorland Publishing.
British Railways Pre-Grouping Atlas. Conolly. Ian Allan.
Great Locomotives of the LNER. Nock. PSL.
Gresley Locomotives. Haresnape. Ian Allan.
LNER Locomotive Allocations. Yeadon. Irwell.
Locomotives of Sir Nigel Gresley. Nock. Railway Publishing.
Locomotives of the LNER. Nock. LNER Publications.
Locomotives of the LNER. RCTS.

Nigel Gresley — Locomotive Engineer. Brown. Ian Allan.
Preserved Locomotives of British Railways. Fox/Webster. Platform Five.
The British Steam Railway Locomotive 1825 – 1925. Ahrons. LPC.
The British Steam Railway Locomotive 1925 – 1965. Nock. Ian Allan.
The Great Eastern Railway. Allen. Ian Allan.
The Gresley Influence. Hughes. Ian Allan.
The West Highland Railway. Thomas. Pan.
Assorted Public and Working Timetables.

Much information has also been gathered from a collection of old and new periodicals. These include:

Steam Classic, Railways, Railway World, Railway Magazine, Trains Illustrated and *Leek Growing in Percy Main*.